SpringerBriefs in Physics

W0080391

For further volumes:
http://www.springer.com/series/8902

Andrei Stalmashonak · Gerhard Seifert
Amin Abdolvand

Ultra-Short Pulsed Laser Engineered Metal–Glass Nanocomposites

 Springer

Andrei Stalmashonak
CODIXX AG
Barleben
Germany

Gerhard Seifert
Centre of Innovation Competence
  SiLi-nano
Martin-Luther-Universität
  Halle-Wittenberg
Halle
Germany

Amin Abdolvand
School of Engineering, Physics
  and Mathematics
University of Dundee
Dundee
UK

ISSN 2191-5423   ISSN 2191-5431   (electronic)
ISBN 978-3-319-00436-5   ISBN 978-3-319-00437-2   (eBook)
DOI 10.1007/978-3-319-00437-2
Springer Cham Heidelberg New York Dordrecht London

Library of Congress Control Number: 2013938036

Printed on acid-free paper

Springer is part of Springer Science+Business Media (www.springer.com)

Foreword

Glasses containing embedded metallic nanoparticles (metal–glass nanocomposites) exhibit peculiar linear and nonlinear optical properties, mainly due to the surface plasmon resonances (SPRs) of the metallic inclusions. The nanoparticles' shapes predominantly and characteristically determine the spectral positions and polarization dependence of the SPRs in the visible and near infrared. The focus of this brief is on the interaction of intense ultra-short (femtosecond) laser pulses with silver nanoparticles embedded in soda-lime glass, and the resulting nanostructural modifications. In particular, and in order to provide a comprehensive physical picture of the processes leading to the laser-induced persistent shape transformation of the nanoparticles, series of experimental results investigating the dependences of the nanoparticles' laser-assisted shape modifications on laser pulse intensity, excitation wavelength, and temperature are considered. The resulting local optical dichroism allows fabrication of flexibly polarizing optical (sub-) microstructures with well-specified optical properties. The recently achieved considerable progress toward fabrication of an all-optical data storage and readout technique based on metal–glass nanocomposites is also discussed.

Preface

From medieval stained glass windows toward all-optical computer chips, an understanding of the interaction of light with a complex dielectric medium has been a key to designing and tailoring the optical properties of photonic devices.

For many centuries, the presence of metal nanoparticles has been evident because of the unusual colour effects associated with them. The red and yellow colours of many medieval church windows originated from silver, gold, and copper nanoparticles embedded in the window glass. The first evidence of using gold nanoparticles in antiquity dates back to the fourth century AD (the Lycurgus Cup). The physics of the processes remained a mystery until Michael Faraday, the well-known nineteenth century physicist, discovered that this effect is due to a new type of optical absorption in metal particles with dimensions substantially less than the wavelength of light.

Metal particles with sizes of the order of one to several hundreds of nanometres are the subject of intensive research efforts across the world. This is due to the fascinating differences in the optical properties they exhibit compared to bulk metals. When a metal particle is smaller than the wavelength of light, the light reflected from it is replaced by light scattering, which is particularly strong at the resonance frequencies of collective electron excitations in the particle. These oscillations are known as the particle's plasmons or surface plasmon resonances (SPRs). For noble and alkali metals, where the conduction electrons are sufficiently free-electron-like, the collective excitations show themselves as pronounced resonance effects in optical scattering and absorption spectra.

Glasses containing embedded metallic nanoparticles exhibit very promising linear and nonlinear optical properties, mainly due to the SPRs of the nanoparticles. The focus of this brief will be on the interaction of intense ultra-short laser pulses with glass nanocomposites comprising silver nanoparticles embedded in soda-lime glass, and nanostructural modifications in metal–glass nanocomposites induced by such laser pulses. In order to provide a comprehensive physical picture of the processes leading to laser-induced persistent shape transformation of the nanoparticles, series of experimental results investigating the dependences of laser-assisted shape modifications of nanoparticles with laser pulse intensity, excitation wavelength, and temperature are considered. In addition, the resulting local optical dichroism allows production of very flexibly polarizing optical (sub-)

microstructures with well-specified optical properties. The achieved considerable progress towards technological application of this technique, in particular for long-term optical data storage, will also be discussed. It is argued that the latter could be utilized for multi-bit encoding in spot sizes down to the diffraction limit, where the information can be read out very fast by wavelength- and polarization-sensitive detection of the transmitted light. The storage capacity of the proposed technique is comparable with that of Blu-ray disks.

Acknowledgments

The authors are very grateful to CODIXX AG for providing glass samples containing Ag nanoparticles for the experiments. We owe many thanks to Prof. Dr. H. Graener for his continuous support, encouragement, and enthusiasm for this and other related work. We would also like to thank Dr. A. Podlipensky, Dr. A. A. Unal, and Dr. M. Kaempfe for fruitful discussions and their contribution to many stages of the work and to Dr. A. C. Hourd for proof reading the text. We express our thanks to Dr. J. Lange, I. Otten and C. Seidel for their technical support. We thank Prof. Dr. W. Hergert, Dr. O. Kiriyenko, and C. Matyssek for providing material on the Electric Field Enhancement and some of the calculations presented in this brief. We extend our cordial thanks to F. Syrovatka (IWZ Materialwissenschaft, MLU Halle), Dr. P. Miclea (MLU, Halle) and Dr. H. Hofmeister (Max Planck Institute of Microstructure Physics-Halle) for help with the scanning and transmission electron microscopy. Last but not least, the substantial financial support from the Deutsche Forschungsgemeinschaft (SFB 418), and also support from the Engineering and Physical Research Council (EPSRC) of the United Kingdom (EP/I004173/1) are gratefully acknowledged. Amin Abdolvand is currently an EPSRC Career Acceleration Fellow at the University of Dundee.

Contents

Chapter 1
Introduction

Glasses and other dielectrics containing metallic nanoparticles are very promising materials for applications in optoelectronics due to their unique linear and non-linear optical properties. These properties are dominated by the strong surface plasmon resonance (SPR) of the metal nanoparticles. The SPR occurs when the electron and light waves couple with each other at a metal-dielectric interface. These are regarded as the collective oscillation of the nanoparticle (NP) electrons. The spectral position of the SPR in the compound materials can be designed within a wide spectral range, from visible to near-infrared, by choice of the electronic properties of the metal and the dielectric matrix [1, 2], or by manipulation of size [1, 3], shape [3, 4], and spatial distribution [5] of the metal clusters. This makes the composite materials attractive for some applications in the field of photonics [6, 7]. One of the main issues in this context is to structure the optical properties of such materials on a micro- or even sub-micron scale. This aspect, in fact, occupies many researchers within the scientific community. The scientific interest comprises the study of optical sub-wavelength structures such as plasmonic waveguides based on metal nanoparticles [8, 9], as well as larger (micrometer) scale optical structures; for the latter nanocomposite materials are very appropriate for production of a number of standard and advanced optical elements such as: gratings, segmented filters and polarizers.

Recently, it was shown that laser-induced techniques represent a very powerful and flexible tool for (local) structuring of the optical properties of metal-doped composite materials [10–18]. Particularly, it was discovered that initially spherical silver nanoparticles embedded in soda-lime glass experience a persisting transformation of their shape when irradiated with intense ultra-short (femtosecond) laser pulses operating near to their SPR band [14–18]. The results can be macroscopically observed as optical dichroism—polarization dependent colour changes.

Our intention here is to provide a comprehensive overview of the various physical processes responsible for permanent nanostructural alterations in glass with embedded silver nanoparticles upon irradiation with intense, ultra-short laser pulses. The studies reported here were aiming to understand the details of the mechanism of NP shape transformation, and to apply this knowledge to achieve

A. Stalmashonak et al., *Ultra-Short Pulsed Laser Engineered Metal–Glass Nanocomposites*, SpringerBriefs in Physics, DOI: 10.1007/978-3-319-00437-2_1, © The Author(s) 2013

optimized optical parameters such as polarization contrast or broadening of the tuning range of dichroism or optical storage of information in nanocomposites. The latter is a necessary prerequisite to exploit the full range of potential applications of this novel technology.

For the sake of conciseness, we do not provide a comprehensive review of the optical properties of nanocomposites with metal particles, which can be found elsewhere [1, 2, 19], but restrict the discussion in Chap. 2 to the most important properties needed for easy comprehension of this review. Chapter 3 will then provide an overview of the possible processes initiated by interaction of ultra-short laser pulses with metal nanoparticles. Taking into account experimental results, as well as numerical temperature modeling and semi-quantitative descriptions of electron and ion emission into the glass matrix, we arrive at a model for the permanent structural change, which facilitates a self-consistent description of all experimental observations. Chapters 4, 5 and 6 cover the influence of various parameters such as irradiated number of laser pulses, laser pulse intensity, excitation wavelength and temperature on the achieved dichroism in the nanocomposite—mainly governed by the change in the NPs' aspect ratio. Based on the conclusions drawn from these studies we will present a simple technique for the fabrication of polarizing optical micro-structures with unprecedented properties, and make the first steps towards fabrication of high-capacity media for optical storage of information. These are presented in Chap. 7.

References

1. Kreibig, U., Vollmer, M.: Optical Properties of Metal Clusters, Springer Series in Material Science, vol. 25. Springer, Berlin (1995)
2. Shalaev, V.M., Kawata, S.: Nanophotonics with surface plasmons. In: Advances in Nano-Optics and Nano-Photonics. Elsevier, UK (2007)
3. Kelly, K.L., Coronado, E., Zhao, L.L., Schatz, G.C.: The optical properties of metal nanoparticles: the influence of size shape and dielectric environment. J. Phys. Chem. B **107**, 668–677 (2003)
4. Jin, R., Cao, Y.C., Hao, E., Merraux, G.S., Schatz, G.C., Mirkin, C.A.: Controlling anisotropic nanoparticle growth through plasmon excitation. Nature **425**, 487–490 (2003)
5. Gudiksen, M.S., Lauhon, L.J., Wang, J., Smith, D.C., Lieber, C.M.: Growth of nanowire superlattice structures for nanoscale photonics and electronics. Nature **415**, 617–620 (2002)
6. Chakraborty, P.: Metal nanoclusters in glasses as non-linear photonic materials. J. Mater. Sci. **33**, 2235–2249 (1998)
7. Gonella, F., Mazzoldi, P.: Handbook of Nanostructured Materials and Nanotechnology, vol. 4. Academic Press, San Diego (2000)
8. Krenn, J.R.: Nanoparticle waveguides: Watching energy transfer. Nature Mater. **2**, 210–211 (2003)
9. Maier, S.A., Kik, P.G., Atwater, H.A., Meltzer, S., Harel, E., Koel, B., Requicha, A.G.: Local detection of electromagnetic energy transport below the diffraction limit in metal nanoparticle plasmon waveguides. Nature Mater. **2**, 229–232 (2003)
10. Wenzel, T., Bosbach, J., Goldmann, A., Stietz, F., Trager, F.: Shaping nanoparticles and their optical spectra with photons. Appl. Phys. B **69**, 513–517 (1999)

11. Stietz, F.: Laser manipulation of the size and shape of supported nanoparticles. Appl. Phys. A **72**, 381–394 (2001)
12. Stepanov, A.L., Hole, D.E., Bukharaev, A.A., Townsend, P.D., Nurgazizov, N.I.: Reduction of the size of the implanted silver nanoparticles in float glass during excimer laser annealing. Appl. Surf. Sci. **136**, 298–305 (1998)
13. Shimotsuma, Y., Yuasa, T., Homma, H., Sakakura, M., Nakao, A., Miura, K., Hirao, K., Kawasaki, M., Qiu, J., Kazansky, P.G.: Photoconversion of Copper Flakes to Nanowires with Ultrashort Pulse Laser Irradiation. Chem. Mater. **19**, 1206–1208 (2007)
14. Kaempfe, M., Rainer, T., Berg, K.-J., Seifert, G., Graener. H.: Ultrashort laser pulse induced deformation of silver nanoparticles in glass. Appl. Phys. Lett. **74**, 1200–1202 (1999)
15. Kaempfe, M., Seifert, G., Berg, K.-J., Hofmeister, H., Graener. H.: Polarization dependence of the permanent deformation of silver nanoparticles in glass by ultrashort laser pulses. Eur. Phys. J. D **16**, 237–240 (2001)
16. Seifert, G., Kaempfe, M., Berg, K.-J., Graener, H.: Production of "dichroitic" diffraction gratings in glasses containing silver nanoparticles via particle deformation with ultrashort laser pulses. Appl. Phys. B **73**, 355–359 (2001)
17. Podlipensky, A., Abdolvand, A., Seifert, G., Graener, H.: Femtosecond laser assisted production of dichroic 3D structures in composite glass containing Ag nanoparticles. Appl. Phys. A **80**, 1647–1652 (2005)
18. Podlipensky, A.V.: Laser assisted modification of optical and structural properties of composite glass with silver nanoparticles. Ph.D. Thesis, Martin-Luther-Universität Halle-Wittenberg. http://sundoc.bibliothek.uni-halle.de/dissonline/05/05H084/t1.pdf (2005)
19. Shalaev, V.M.: Optical Properties of Nanostructured Random Media, Topics in Applied Physics, vol. 82. Springer, Berlin, DE (2002)

Chapter 2
Optical Properties of Nanocomposites Containing Metal Nanoparticles

Interaction of light with nanocomposites reveals novel optical phenomena indicating unrivalled optical properties of these materials. The linear and non-linear optical response of metal nanoparticles is specified by oscillations of the surface electrons in the Coulomb potential formed by the positively charged ionic core. This type of excitation is called the Surface Plasmon (SP). In 1908 Mie [1] proposed a solution of Maxwell's equations for spherical particles interacting with plane electromagnetic waves, which explains the origin of surface plasmon reso-nance (SPR) in the extinction spectra and colouration of metal colloids.

During the last century optical properties of nanoparticles have extensively been studied and metal-dielectric nanocomposites have found various applications in different fields of science and technology [2–6]. Since the optical properties of metal nanoparticles are governed by SPR, they are strongly dependent on the nanoparticles' size, shape, concentration and spatial distribution as well as on the properties of the surrounding matrix. Control over these parameters enables such metal-dielectric nanocomposites to become promising media for development of novel non-linear materials, nanodevices and optical elements.

In this section the SPR and main optical properties of metal nanoparticles embedded in a dielectric medium will be considered. A comprehensive review of the optical properties of nanostructured random media can be found in Refs. [7, 8].

2.1 Surface Plasmon Resonance of Isolated Metal Nanoparticles

Exposure of a metal nanoparticle to an electric field results in a shift of the free conduction electrons with respect to the particle's metal ion-lattice. The resulting surface charges of opposite sign on the opposite surface elements of the particles (see Fig. 2.1) produce a restoring local field within the nanoparticle, which rises with the increasing shift of the electron gas relative to the ionic background. The coherently shifted electrons of the metal particle together with the restoring field

A. Stalmashonak et al., *Ultra-Short Pulsed Laser Engineered Metal–Glass Nanocomposites*, SpringerBriefs in Physics, DOI: 10.1007/978-3-319-00437-2_2, © The Author(s) 2013

Fig. 2.1 Plasmon
oscillations in metal spheres
induced by an
electromagnetic wave

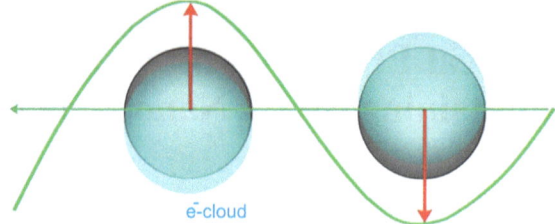

ē-cloud

consequently represent an oscillator, whose behavior is defined by the electron
density and the geometry of the particle. Throughout this text the nanoparticles'
resonances are called surface plasmons on metal nanoparticles.

An exact analytical, theoretical description of SPs of spherical metal nanoparticles is part of Mie's theory for scattering and absorption of light by spheres
[7, 9]. According to the theory, different eigenmodes of the spherical particles are
dipolar or multipolar in character. For particles that are small compared to the
local variations of the involved electromagnetic fields, the quasi-static approximation is valid [7]. It assumes the exciting field to be homogeneous and not
retarded over the particle's volume. Under these assumptions, the results of
electrostatics can be applied by using the corresponding frequency dependent
dielectric function. In this case, the polarizability α and induced dipole moment
p of a metal sphere embedded in dielectric are given as [10]:

$$\alpha = 4\pi R^3 \frac{\varepsilon_i(\omega) - \varepsilon_h}{\varepsilon_i(\omega) + 2\varepsilon_h}, \tag{2.1}$$

$$\vec{p}(\omega) = \alpha\varepsilon_0\vec{E}_0(\omega) = 4\pi\varepsilon_0 R^3 \frac{\varepsilon_i(\omega) - \varepsilon_h}{\varepsilon_i(\omega) + 2\varepsilon_h} \vec{E}_0(\omega), \tag{2.2}$$

where R is the radius of the nanoparticle, E_0 the electric field strength of the
incident electromagnetic wave, ε_0 the electric permittivity of vacuum, and $\varepsilon_i(\omega)$
and ε_h are the relative complex electric permittivity of the metal and host matrix
respectively.

The absorption cross-section of a spherical metal inclusion placed in a transparent dielectric matrix, where the imaginary part of the relative complex electric
permittivity approaches zero ($\mathrm{Im}[\varepsilon_h]{\to}0$), is then given as:

$$\sigma(\omega) = 12\pi R^3 \frac{\omega}{c} \varepsilon_h^{3/2} \frac{\varepsilon_i''(\omega)}{[\varepsilon_i'(\omega) + 2\varepsilon_h]^2 + \varepsilon_i''(\omega)^2} \tag{2.3}$$

where $\varepsilon_i'(\omega)$ and $\varepsilon_i''(\omega)$ are the real and imaginary parts of the electric permittivity
of the metal, which in turn can be described by the Drude-Sommerfeld formula:

$$\varepsilon_i(\omega) = \varepsilon_b + 1 - \frac{\omega_p^2}{\omega^2 + i\gamma\omega}. \tag{2.4}$$

Here, γ is the damping constant of the electron oscillations and ε_b is the complex electric permittivity associated with interband transitions of the core electrons in the atom. The free electron plasma frequency is given by:

$$\omega_p = \sqrt{\frac{Ne^2}{m\varepsilon_0}}, \tag{2.5}$$

where N is the density of the free electrons and m is the effective mass of an electron.

As can be seen from Eqs. 2.1–2.3, the well-known Mie resonance occurs at the SP frequency ω_{SP} under the following conditions:

$$\left[\varepsilon_i'(\omega) + 2\varepsilon_h\right]^2 + \varepsilon_i''(\omega)^2 \rightarrow Minimum. \tag{2.6}$$

If the imaginary part of the metal electric permittivity is small in comparison with $\varepsilon_i'(\omega)$, or has small frequency dependence, then Eq. 2.6 can be written as:

$$\varepsilon_i'(\omega_{SP}) = -2\varepsilon_h. \tag{2.7}$$

Thus, if the condition represented in Eq. 2.7 is fulfilled, the dipole moment and local electric field in the vicinity of the nanosphere grow resonantly and can achieve magnitudes enhanced by many orders, overcoming the field of the incident wave. This phenomenon is responsible for the SP enhanced non-linearities in metal colloids.

Equation 2.7 requires the real part of the dielectric function of metals to be negative. This is indeed the case for noble metals in the visible spectral region (as an example see Fig. 2.2 for Ag). For a silver nanoparticle surrounded by a dielectric environment with $\varepsilon_h = 2.25$, the resonance condition is observed to occur at around 400 nm (Fig. 2.3). This results in the observation of bright colours both in transmitted and reflected light from such a medium.

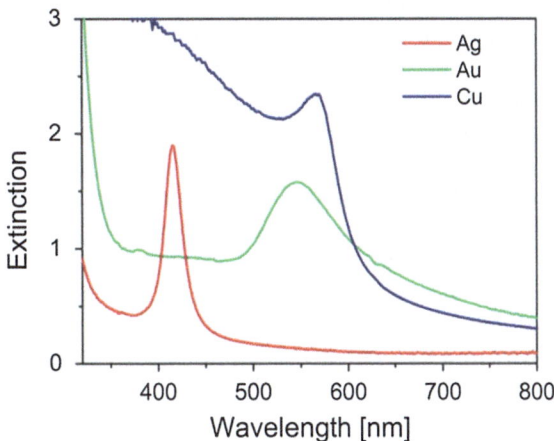

Fig. 2.2 Extinction spectra of glass containing *spherical silver, gold* and *copper* nanoparticles

Fig. 2.3 The dielectric function of silver (*solid lines* data adapted from Ref. [11]): comparing its real part to $-2\varepsilon_m$ (*dashed line*) of an idealized dielectric medium with a wavelength independent dielectric constant of $\varepsilon_m = 2.25$ ($n = 1.5$), the surface plasmon resonance condition is found at ~ 410 nm (*vertical arrow*)

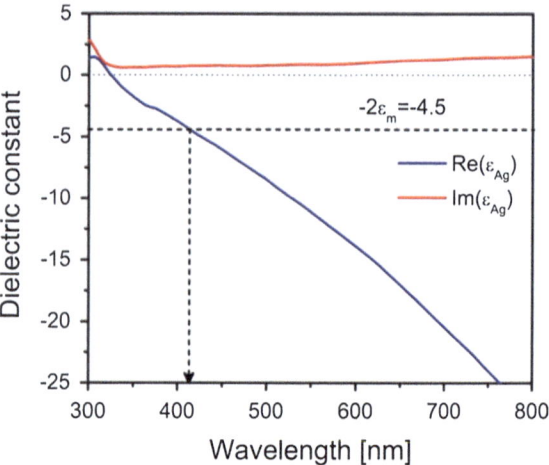

Using Eq. 2.7 and by substituting the real part of the metal electric permittivity from Eq. 2.4, the position of the SP resonance can be expressed as follows:

$$\omega_{SP}^2 = \frac{\omega_p^2}{\text{Re}[\varepsilon_b] + 1 + 2\varepsilon_h} - \gamma^2. \qquad (2.8)$$

Core electrons have a significant influence on the SP and define the position of the SPR in the extinction spectra (Fig. 2.2) for different noble metals. For instance, silver nanoparticles embedded in glass matrix exhibit a SP band at about 415 nm. In turn, SP for Au and Cu nanoparticles is shifted in the red spectral range and peaked at 528 and 570 nm, respectively. The broad absorption bands below 500 nm for both Au- and Cu-containing nanocomposite glasses are associated with interband transitions, namely from the d- to s-shell, of the core electrons in the metal atoms. However, for silver the interband resonance is peaked at 310 nm (4 eV), far away from the SP resonance [11].

On the other hand, Eq. 2.8 qualitatively describes a dependence of the SP resonance on the dielectric properties of the host matrix, which the metal nanoparticles are incorporated in. An increase of dielectric constant (refractive index) evokes a shift of absorption maximum towards longer wavelengths [7, 12, 13] (as would be expected from Fig. 2.3). Figure 2.4 represents the spectral positions of SPRs of silver nanoparticles embedded in vacuum ($\varepsilon_h = 1$) and glass ($\varepsilon_h = 2.25$). It is clearly seen that the SP resonance maxima are more red-shifted for nanocomposites having a matrix with higher dielectric constant.

Figure 2.4 also shows that the position of the SPR depends on the size of metallic nanoparticles. In fact, its position remains quasi-constant for nanoparticles with radii smaller than 10–15 nm, while the band's half-width for these clusters differs by a factor of 4. This is often described as an intrinsic size effect [7, 14, 15]. If the particle size is below the dimension of the mean free path of the electrons in the metal (≈ 10–15 nm) [16], the electron scattering at the particle surface mainly

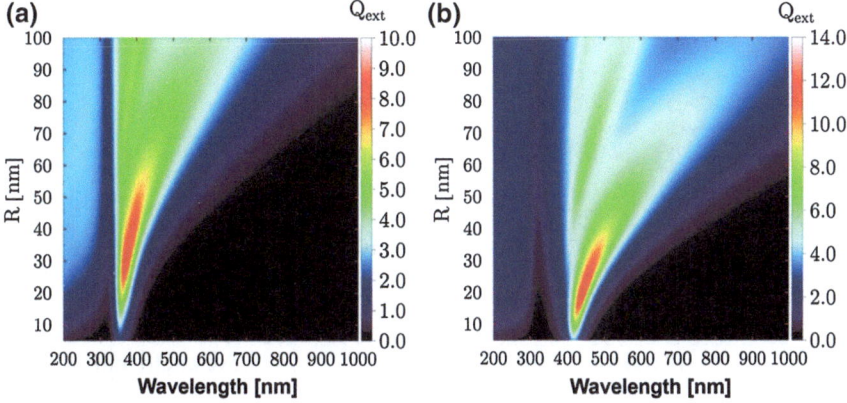

Fig. 2.4 Extinction spectra of spherical silver nanoparticles in **a** vacuum and **b** glass as a function of particle size

increases the imaginary part of the dielectric function. For the smaller particles (>1 nm) the spill-out of electrons from the particle surface should be taken into account, which results in an inhomogeneous dielectric function. As a result of this effect, very broad plasmon bands are observed for small nanoparticles (not included in Fig. 2.4).

The SPR shifts towards longer wavelengths with a simultaneous increase in the band half-width for nanospheres with radii larger than 15 nm (Fig. 2.4). This effect for the larger particle is referred to as an extrinsic size effect [7, 14, 17–19]. In this case, higher-order (such as quadrupolar) oscillations of conduction electrons become important.

From the size dependence of the SP, it is quite obvious that metal nanoparticles with non-spherical shape will show several SP resonances in their spectra. For instance, ellipsoidal particles with axes a ≠ b ≠ c own three SP modes corresponding to polarizabilities along the principal axes given as:

$$\alpha_k(\omega) = \frac{4\pi}{3}abc\frac{\varepsilon_i(\omega) - \varepsilon_h}{\varepsilon_h + (\varepsilon_i(\omega) + \varepsilon_h)L_k}, \tag{2.9}$$

where L_k is the geometrical depolarization factor for each axis ($\Sigma L_k = 1$). Moreover, an increase in the axis length leads to the minimization of the depolarization factor. For a spherical particle $L_a = L_b = L_c = 1/3$.

Thus, if the propagation direction and polarization of the electromagnetic wave do not coincide with the axes of the ellipsoid, the extinction spectra can demonstrate three separate SP bands corresponding to the oscillations of the free electrons along these axes [7]. For spheroids one has: $a \neq b = c$, and the spectra exhibit *two* SP resonances. However, if the incident light is polarized parallel to one of the axes, only one single SP band corresponding to the appropriate axis is observed (Fig. 2.5). The band lying at higher wavelengths is referred to as the long axis, while the small axis demonstrates resonance at shorter wavelengths compared

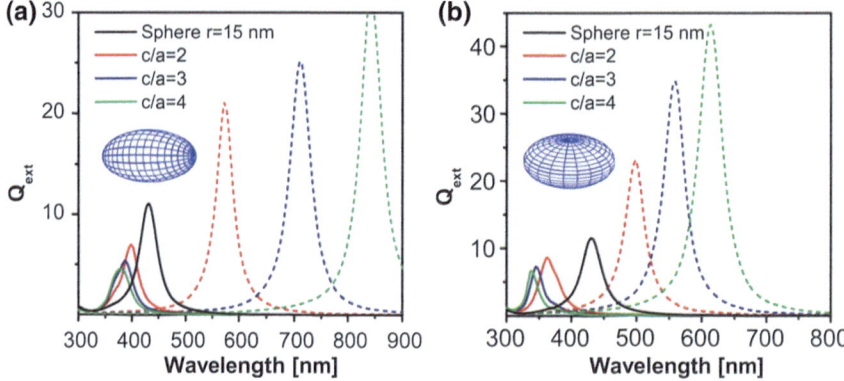

Fig. 2.5 Calculated using the Mie theory for spheroids [22], polarized extinction spectra are shown for **a** prolate and **b** oblate silver spheroids with different aspect ratios, which are embedded in glass. The volume of the spheroids is equal to the volume of a nanosphere with radius of 15 nm. *Dashed curves* polarization of the light is parallel to the long axis; *solid lines* parallel to the short axis. The insets schematically show the shape of the spheroids

to the single resonance of a nanosphere of the same volume. The spectral separation of the two surface plasmon bands of the ellipsoidal nanoparticle strongly depends on its aspect ratio [20, 21], which is defined as the ratio of the long to the short axes. At the same time, it is clearly seen that for prolate and oblate spheroids having the same aspect ratio, the positions of SP resonances are different. Namely, the spectral separation between SP bands is higher for the nanoparticles having a zeppelin-like shape.

For many years now, the dichroic property of elongated metallic nanoparticles has been used for manufacture of broad-band high-contrast polarizers [22]. This became possible because the position of the SP resonance can be designed within a broad spectral range by an appropriate choice of aspect ratio between the axes of the nanoparticles. This aspect will be discussed in more detail in the next sections.

2.2 Optical Properties of Nanocomposites with a High Fraction of Metal Nanoparticles

Increasing the volume fraction of metal nanoparticles in a medium leads to a decrease in the average inter-particle distances. Thus, enhancement of the dipole moment of spherical metal NPs by excitation near to their SP resonance results in strong collective dipolar interactions between the nanoparticles, which in turn affect the linear and non-linear optical properties of the nanocomposite material. For the purpose of this work it is sufficient to describe these effects using the approximation of the well-known Maxwell–Garnett effective medium theory. This theory is widely applied to describe the optical properties of metal particles in dielectric matrices [8, 9, 23, 24]. The theory does not correctly take into account the multipolar interactions between nanoparticles considered in other works [25, 26]. However, it describes quite well the position and shape of the SP resonance and its dependence on the metal filling factor [8].

The effective dielectric constant $\varepsilon_{eff}(\omega)$ of a composite material with spherical metal inclusions having a filling factor f (volume of the silver inclusions per unit volume of the composite material $f = V_{Ag}/V_{total}$) is given by the expression:

$$\varepsilon_{eff}(\omega) = \varepsilon_h \frac{(\varepsilon_i(\omega) + 2\varepsilon_h) + 2f(\varepsilon_i(\omega) - \varepsilon_h)}{(\varepsilon_i(\omega) + 2\varepsilon_h) - f(\varepsilon_i(\omega) - \varepsilon_h)}, \tag{2.10}$$

where $\varepsilon_i(\omega)$ and ε_h are the complex electric permittivities of the metal (given by the Eq. 2.4) and the host matrix, respectively. Based on this description, the complex index of refraction of a composite medium can be defined as

$$n(\omega) = n' + in'' = \sqrt{\varepsilon_{eff}(\omega)}. \tag{2.11}$$

Hence, the absorption coefficient α and refractive index n' of the medium with dielectric constant $\varepsilon_{eff}(\omega)$ can be expressed as

$$\alpha = \frac{2\omega}{c} \operatorname{Im}\sqrt{\varepsilon_{eff}(\omega)}, \tag{2.12}$$

$$n'(\omega) = \operatorname{Re}\sqrt{\varepsilon_{eff}(\omega)}, \tag{2.13}$$

where c is the light velocity. Using Eqs. 2.10–2.13, the absorption cross-section and dispersion spectra (Fig. 2.6a, b) of glass containing spherical silver nanoparticles can be calculated as a function of the volume filling factor of the metal clusters in the glass matrix, using: $\varepsilon_h = 2.3$, $\omega_p = 9.2$ eV, $\gamma = 0.5$ eV [27], $\varepsilon_b = 4.2$ [24].

The collective dipolar interactions between nanoparticles cause a broadening and red-shift of the absorption band with increasing filling factor of the inclusions in the glass matrix (Fig. 2.6a). The effective refractive index of the composite glass also changes with growing filling factor (Fig. 2.6b)—at low content of silver nanoparticles in glass ($f = 0.001$) the refractive index is identical to that of clear glass ($n' = 1.52$), higher filling factors result in significant modifications of the dispersion spectra. For $f = 0.1$, the refractive index varies between ~ 1.2 and 2.1 on different sides of the SP resonance. As shown in Fig. 2.6c, the reflectivity R—given for normal incidence by:

$$R(\omega) = \left| \frac{n(\omega) - 1}{n(\omega) + 1} \right|^2 \tag{2.14}$$

Fig. 2.6 a Absorption cross-section, **b** dispersion and **c** reflection spectra of composite glass containing Ag nanoparticles calculated according to the Maxwell–Garnett theory

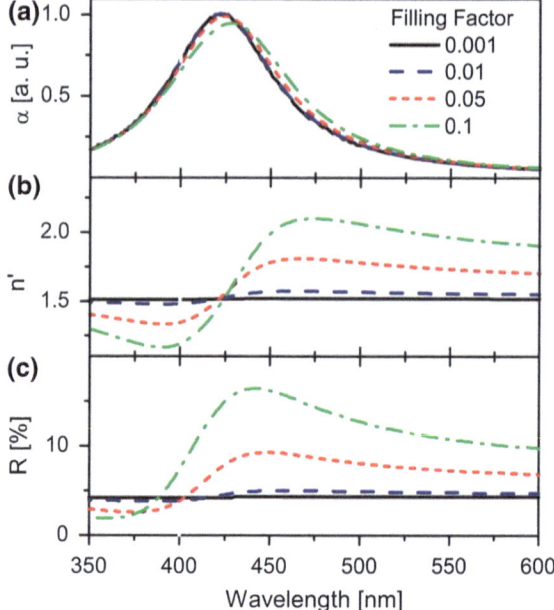

also changes upon increasing the volume filling factor. In particular, in the visible range the main effect is an increase in reflectivity of the composite medium with an increasing content of nanoparticles.

2.3 Preparation and Characterization of Glass Samples Containing Silver Nanoparticles

The samples were prepared from soda-lime float glass (72.5 SiO_2, 14.4 Na_2O, 0.7 K_2O, 6.1 CaO, 4.0 MgO, 1.5 Al_2O_3, 0.1 Fe_2O_3, 0.1 MnO, 0.4 SO_3 in wt %) by Ag^+–Na^+ ion exchange. For the ion exchange process the glass substrate is placed in a mixed melt of $AgNO_3$ and KNO_3 at 400 °C [12, 28]. The thickness of the glass substrate, time of the ion exchange process and weight concentration of $AgNO_3$ in the melt determine the concentration and distribution of Ag^+ ions in the glass. Subsequent thermal annealing of the ion exchanged glass in an H_2 reduction atmosphere, typically at 400–450 °C, results in the formation of spherical silver NPs [12]. As could be expected, size and depth distribution of the Ag NPs in the glass sample strongly depend on the temperature and duration of the Na–Ag ion exchange as well as on the duration of the annealing. In our case, the spherical Ag NPs of 30–40 nm in mean diameter (Fig. 2.7a) are distributed in a thin surface layer of approximately 6 µm thickness (with total thickness of the glass plate being 1 mm).

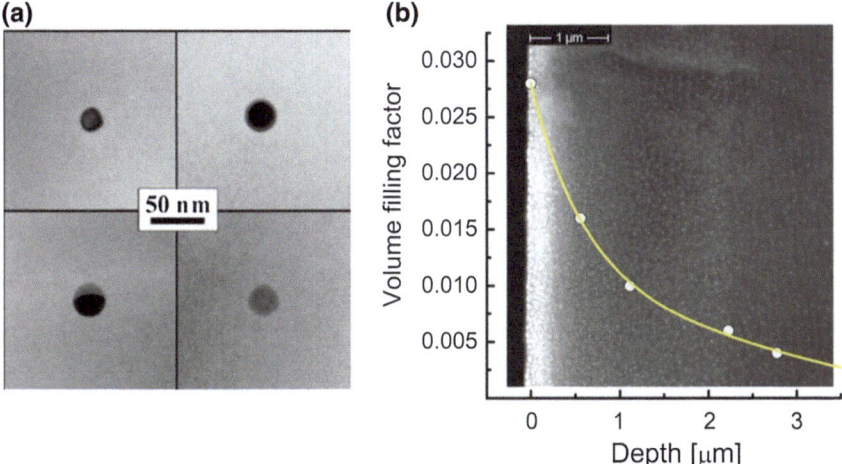

Fig. 2.7 **a** TEM image of typical spherical silver nanoparticles in nanocomposite glass. **b** SEM image of the cross-section of a glass sample containing spherical silver nanoparticles (Ag particles are reproduced as *white spots*). The gradient of the volume filling factor of Ag nanoparticles is shown superimposed (the x-axis was adjusted to the length scale of the image)

The scanning electron microscopy (SEM) image of the cross-section of the nanocomposite is shown in Fig. 2.7b, where silver particles are reproduced as white spots. In order to obtain information on the depth distribution of silver NPs in the glass, surface layers of various thicknesses were removed from the sample by etching in 12 % HF acid for different retention times. After this procedure SEM images were recorded for all etched surfaces [e.g., Fig. 2.8a, increasing etching time from (i) to (iv)], as well as optical extinction spectra (see Fig. 2.8b). The area fraction of silver derived from the SEM images was then converted to a volume fill factor assuming a typical electron penetration depth of 500 nm. The result is given as the superimposed curve in Fig. 2.7b, showing the highest silver content of $f = 0.028$ directly below the glass surface. Within a few micrometers the fill factor then strongly decreases with increasing distance from the surface.

Figure 2.8b depicts the corresponding extinction spectra with the same lettering as in Fig. 2.8a. However, it should be noticed that the optical spectra integrate over the whole particle-containing layer. Thus, for the original sample the absorption around SP resonance is very high, discouraging any detailed analysis of the spectral band shape. The same holds for extinction after the shortest etching time [Fig. 2.8b, curve (i)]; however, at least one can estimate for this case an extinction peak wavelength in the range of 420–440 nm. Further etching of the sample results in a fading of the extinction band caused by the decrease in thickness of the silver-nanoparticle containing layer. Spectrum (i) indicates that the uppermost metal-rich layers are responsible for the shift observed in the red wing of the SP band towards longer wavelengths.

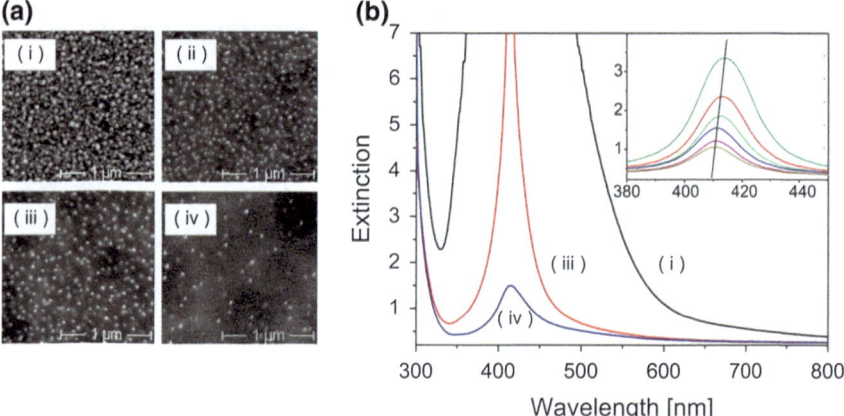

Fig. 2.8 **a** SEM images of etched samples showing Ag nanoparticles (volume fill factor: *i* 0.01, *ii* 0.006, *iii* 0.004, *iv* 0.001). **b** Extinction spectra of samples with spherical silver nanoparticles after different times of etching in 12 % HF acid. Labeling of the spectra is according to the SEM images shown in (**a**). The samples with lower fill factor exhibit lower extinction

Etching the samples for longer times, so that one ends up at a residual filling factor of less than 0.004, leads to an easily measurable extinction spectrum. The corresponding evolution of the spectra with etching time is shown in more detail in the inset of Fig. 2.8b. It is seen that a decrease of filling factor (by longer time of etching) leads to a slight shift of the SP band maxima to shorter wavelengths. This is compatible with the Maxwell–Garnett theory, which predicts a red-shift of the SP band for the samples with higher filling factors [7, 8, 29].

It should be mentioned here that for the study of the basic physical processes of nanoparticle shape transformation, etched samples with a maximum Ag filling factor of 10^{-3} were used, where the NPs are, to good approximation, understood as non-interacting (isolated) nanoparticles. However, the experiments related to the maximization of polarization contrast were performed on samples with considerably higher filling factor ($f \sim 0.01$).

References

1. Mie, G.: Beiträge zur Optik trüber Medien, speziell kolloidaler Metallösungen. Ann. Phys. **25**, 377–445 (1908)
2. Shalaev, V.M., Kawata, S.: Nanophotonics with surface plasmons. Advances in Nano-Optics and Nano-Photonics. Elsevier, UK (2007)
3. Brongersma, M L.; Kik, P G. Surface Plasmon Nanophotonics; Springer Series in OPTICAL SCIENCES 131; Springer: Berlin, DE, 2007
4. Tominaga, J., Tsai, D.P.: Optical Nanotechnologies: The Manipulation of Surface and Local Plasmons, Topics in Applied Physics, vol. 88. Springer, Berlin (2003)

5. Righini, M., Girard, C., Quidant, R.: Light-induced manipulation with surface plasmons. J. Opt. A: Pure Appl. Opt. **10**, 093001 (2008)
6. Wang, J., Blau, W.J.: Inorganic and hybrid nanostructures for optical limiting. J. Opt. A: Pure Appl. Opt. **11**, 024001 (2009)
7. Kreibig, U., Vollmer, M.: Optical Properties of Metal Clusters, Springer Series in Material Science, vol. 25. Springer, Berlin (1995)
8. Shalaev, V.M.: Optical Properties of Nanostructured Random Media, Topics in Applied Physics, vol. 82. Springer, Berlin (2002)
9. Bohren, C.F., Huffman, D.R.: Absorption and Scattering by Small Particles. Wiley, New York (1983)
10. Maier, S.A.: Plasmonics: Fundamentals and Applications. Springer, Berlin (2007)
11. Kresin, V.V.: Collective resonances in silver clusters: Role of d electrons and the polarizationRfree surface layer. Phys. Rev. B **51**, 1844–1899 (1995)
12. Berg, K.-J., Berger, A., Hofmeister, H.: Small silver particles in glass surface layers produced by sodium-silver ion exchange–their concentration and size depth profile. Z. Phys. D **20**, 309–311 (1991)
13. Hilger, A., Tenfelde, M., Kreibig, U.: Silver nanoparticles deposited on dielectric surfaces. Appl. Phys. B **73**, 361–372 (2001)
14. Link, S., El-Sayed, M.A.: Optical Properties and Ultrafast Dynamics of Metallic Nanocrystals. Ann. Rev. Phys. Chem. **54**, 331–366 (2003)
15. Berciaud, S., Cognet, L., Tamarat, P., Lounis, B.: Observation of Intrinsic Size Effects in the Optical Response of Individual Gold Nanoparticles. Nano Lett. **5**, 515–518 (2005)
16. Seah, M.P., Dench, W.A.: Quantitative electron spectroscopy of surfaces: A standard data base for electron inelastic mean free paths in solids. Surf. Interface Anal. **1**, 2–11 (1979)
17. Genzel, L., Martin, T.P., Kreibig, U.: Dielectric function and plasma resonances of small metal particles. Z. Phys. B **21**, 339–346 (1975)
18. Kreibig, U., Genzel, L.: Optical absorption of small metallic particles. Surf. Sci. **156**, 678–700 (1985)
19. Amendola, V., Meneghetti, M.: Size Evaluation of Gold Nanoparticles by UV–vis Spectroscopy. J. Phys. Chem. C **113**, 4277–4285 (2009)
20. Postendorfer, J.: Numerische Berechnung von Extinktions- und Streuspektren spharoidaler Metallpartikel beliebiger Größe in dielektrischer Matrix. PhD thesis, Martin-Luther University Halle-Wittenberg (1997)
21. Voshchinnikov, N.V., Farafonov, V.G.: Optical properties of spheroidal particles. Astrophys. Space Sci. **204**, 19–86 (1993)
22. See for example the following link: http://www.codixx.de/
23. Marton, J.P., Lemon, J.R.: Optical Properties of Aggregated Metal Systems. Phys. Rev. B **4**, 271–280 (1971)
24. Xu, G., Tazawa, M., Jin, P., Nakao, S.: Surface plasmon resonance of sputtered Ag films: substrate and mass thickness dependence. Appl. Phys. A **80**, 1535–1540 (2005)
25. Markel, V.A., Muratov, L.S., Stockman, M.I., George, T.F.: Theory and numerical simulation of optical properties of fractal clusters. Phys. Rev. B **43**, 8183–8195 (1991)
26. Markel, V.A., Shalaev, V., Stechel, E.B., Kim, W., Armstrong, R.L.: Small-particle composites. I. Linear optical properties. Phys. Rev. B **53**, 2425–2436 (1996)
27. Lamprecht, B., Leitner, A., Ausseneg, F.G.: Femtosecond decay-time measurement of electron-plasma oscillation in nanolithographically designed silver particles. Appl. Phys. B **64**, 269–272 (1997)
28. Hofmeister, H., Drost, W.G., Berger, A.: Oriented prolate silver particles in glass-characteristics of novel dichroic polarizers. Nanostruct. Mater. **12**, 207–210 (1999)
29. Gittleman, J.I., Abeles, B.: Comparison of the effective medium and the Maxwell-Garnett predictions for the dielectric constants of granular metals. Phys. Rev. B **15**, 3273–3275 (1977)

Chapter 3
Interaction of Ultra-Short Laser Pulses with Metal Nanoparticles Incorporated in Dielectric Media

This chapter is dedicated to the general understanding of laser pulse interaction with metal nanoparticles. The aim is to collect all physical processes, which may occur when the laser pulse starts to interact with nanoparticles, as well as all later events and mechanisms, which are triggered by this interaction.

Depending on the laser parameters (e.g., weak and strong excitation regimes) and nanoparticle properties, several physical phenomena may occur. Recent investigations of laser pulse interaction with metal nanoparticles are mostly concentrated on the SP dynamics and the energy exchange (relaxation) mechanisms arising as a result of the interaction (see Refs. [1–6] for a review). These studies employ weak laser pulses to excite the nanoparticles, thereby ensuring only weak electronic perturbations occur. In this low perturbation regime, the changes induced in the SP bands of the nanoparticles are transient and totally reversible. For strong excitation as used in this work, however, the energy absorbed by the nanoparticle becomes very high, which creates large perturbations for the nanoparticle electrons leading to persistent (irreversible) changes to the nanoparticle. In this regime the processes related to the heating and cooling of nanoparticles—e.g., electron–electron (e–e), electron–phonon (e–p) scattering, etc.,—have to be modified. At the same time, this strong excitation can open up additional channels for the energy relaxation in the form of e.g., hot electron and ion emissions (see for example Refs. [7–10]). Although it is currently impossible to account for all the complicated many-body interactions among electrons, phonons, ions etc., some theoretical estimation will be presented.

3.1 Energy Relaxation Following the Excitation of the Nanoparticle: Weak Perturbation Regime

The absorption of a femtosecond (fs) laser pulse produces a coherent collective oscillation of the nanoparticle electrons (Fig. 3.1a). During this quasi-instantaneous process, the phase memory is conserved between the electromagnetic field

A. Stalmashonak et al., *Ultra-Short Pulsed Laser Engineered Metal–Glass Nanocomposites*, SpringerBriefs in Physics, DOI: 10.1007/978-3-319-00437-2_3, © The Author(s) 2013

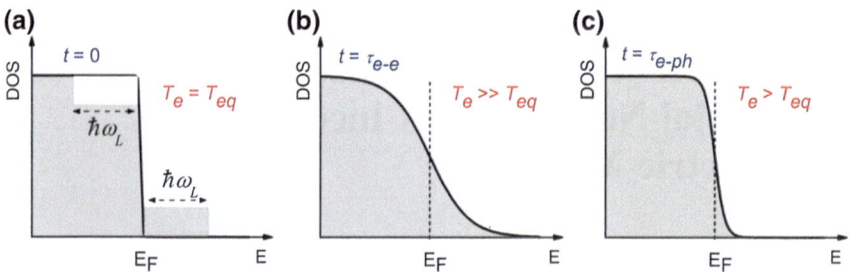

Fig. 3.1 Laser excitation and the subsequent electron dynamics. **a** The electronic distribution is at an equilibrium temperature T_{eq}. Absorption of a laser pulse energy density with photon energy of $\hbar\omega_L$ creates a non-thermal electronic distribution represented by the rectangles. **b** The electrons thermalize to a hot Fermi distribution ($T_e \gg T_{eq}$) through electron–electron scatterings within several hundred femtoseconds. **c** Electrons cool down by sharing their energy with the lattice through electron–phonon coupling processes, reaching a temperature which is equal to the lattice temperature $T_e = T_l > T_{eq}$. DOS stands for the electron density of states

and the electronic states, and the density of excited states depends on the spectral shape of the laser pulse. The corresponding electron distribution is non-thermal [11–13] and lasts for a few femtoseconds [1, 2]. Electrons having energies between $E_F - \hbar\omega$ and E_F are excited above the Fermi energy with final energies between E_F and $E_F + \hbar\omega$. This is depicted in Fig. 3.1a where the excitation is sketched with rectangular-shaped boxes, whose dimensions are determined by the energy of the exciting laser pulse $\hbar\omega$ as the length of the box and the absorbed energy density as the width.

The next step of the energy relaxation corresponds to thermalization of the electrons. The occupied electronic states tend to a Fermi–Dirac distribution with a well-defined temperature, which depends on the laser pulse intensity. The phase coherence is lost and the collective modes have decayed into quasi-particle pairs. Figure 3.1b shows the equilibrated thermal Fermi distribution following the e–e scattering processes. The excited electrons possess high energies above the Fermi level, and the resulting temperature of the electronic system is much higher than the equilibrium temperature before the laser excitation.

Several time-resolved photoemission experiments, performed in noble metal films, have shown that the temporal scale of this thermalization process is a few hundred femtoseconds [12, 14–17]. For small particles, with diameters less than a few tens of nanometers, the scattering time of the electrons at the particle surface is also around a few hundred femtoseconds. Voisin et al. [18] reported an internal electron thermalization time of ≈ 350 fs for 12 nm radius Ag nanoparticles in a BaO-P$_2$O$_5$ matrix, which is comparable to the one determined in Ag films [5].

The time needed for the internal thermalization decreases for smaller nanoparticles, for example, it is around 150 fs for 2 nm radius Ag nanoparticles embedded in an Al$_2$O$_3$ matrix. For the internal electron thermalization of colloidal Au nanoparticles of 9 and 48 nm in radius, decay lifetimes of 500 and 450 fs respectively were found [2]. The size dependence of the thermalization time is in

good agreement with a simple model which phenomenologically introduces surface induced reduction of the Coulomb interaction screening due to the spill-out
and *d*-electron wave function localization effects [18, 19].

Another important mechanism in the electron dynamics, which is shown in
Fig. 3.1c, is the energy transfer to the lattice. The hot electrons cool externally by
electron–phonon (e–p) interactions until the temperatures of the electron gas and
the lattice are equilibrated. The resulting electronic temperature is lower than its
peak value, but higher than the equilibrium temperature. Since the e–p interactions
occur on a timescale comparable with the internal electron thermalization, the e–e
and e–p relaxation cannot be understood as separated processes occurring in
sequential order. This means that the non-thermal electrons of Fig. 3.1a already
interact with the phonons during the same time as they scatter with themselves to
achieve the Fermi distribution of Fig. 3.1b. This simultaneous e–p coupling is an
important channel of electron relaxation, heating the nanoparticle lattice.

There have been many attempts to define the timescale of this thermalization
process in different combinations of metals and matrices [20–27]. Normally one
can expect that the lattice heating needs more time than the heating of electrons,
and that the maximum lattice temperature cannot reach as high as the electron
temperatures since the electronic heat capacity is approximately *two* orders of
magnitude smaller than the lattice heat capacity. Hartland et al. showed that the
thermalization time depends on the laser intensity [8, 28]. In the next section we
shall consider the two-temperature model (2TM, also known as TTM in the literature), which describes the thermal situations of electrons and phonons and the
heat transfer between these two systems. This model will be extended to very high
electronic temperatures in order to account for the conditions of the strong excitation regime.

The last step in the relaxation is the energy transfer to the dielectric matrix. This
transfer corresponds to the heat diffusion from the metal to the environment. It is
therefore sensitive to the thermal conductivity of the surrounding medium and, as
will be shown later, plays an important role in the mechanism of nanoparticle
shape transformation. Therefore, this process will also be considered in the next
sections.

3.2 Electron–Phonon Coupling and Electron Heat Capacity of Silver Under Conditions of Strong Electron–Phonon Non-Equilibrium

Upon pulse interaction with the nanoparticle, the electrons heat up gradually to a
hot electronic distribution. During and after their heating, the electrons couple with
the nanoparticle lattice vibrations (the phonons) and heat up the nanoparticle. The
heat gained by the nanoparticle lattice can be found from the heat lost by the
electrons using the two-temperature model (2TM) [1, 29], where the heat flow

between the two subsystems (electrons and lattice) is defined by two coupled differential equations. 2TM is the commonly accepted theory to describe the energy relaxation mechanisms between electrons and the lattice. The electronic system is characterized by the electron temperature T_e and the phononic system by a lattice temperature T_l. The electron–phonon (e–p) coupling factor $G(T_e)$ is responsible for the energy transfer between the two subsystems. The heat equations describing the temporal evolution of T_e and T_l are given as follows:

$$C_e(T_e)\frac{\partial T_e}{\partial t} = -G(T_e)(T_e - T_l) + S(t),\qquad(3.1)$$

$$C_l\frac{\partial T_l}{\partial t} = G(T_e)(T_e - T_l),\qquad(3.2)$$

where $C_e(T_e)$ and C_l are the electronic and lattice heat capacities, respectively; $S(t)$ in Eq. (3.1) is a source term describing the absorbed laser pulse energy per nanoparticle, which can be given as:

$$S(t) = \frac{I\sigma_{abs}}{V_{NP}} \cdot \exp\left[-4\ln 2 \cdot (t/\tau_{FWHM})^2\right].\qquad(3.3)$$

Here, I is the peak pulse intensity, σ_{abs} is the absorption cross-section of a single nanoparticle ($\approx 3{,}000$ nm^2 for a silver nanoparticle in a dielectric environment with a refractive index of n = 1.52 [30]); V_{NP} is the nanoparticle volume, and τ_{FWHM} determines the full width at half maximum (FWHM) of the temporal pulse profile.

In order to yield a quantitative description of the kinetics of energy redistribution in the irradiated target, it is crucial to use adequate, temperature dependent thermo-physical properties of the target material. Looking at the 2TM equation for the electron temperature (Eq. 3.1), this problem concerns the e–p coupling factor, the electron heat capacity, and the heat conductivity. Due to the small heat capacity of electrons in metals and the finite time needed for the e–p equilibration, irradiation by a short laser pulse can transiently bring the target material to a state of strong electron-lattice non-equilibrium. In this situation, the electron temperature can rise to some ten thousand Kelvins (comparable to the Fermi energy) while the lattice still remains cold. At such high electron temperatures, the thermo-physical properties of the material can be affected by the thermal excitation of the lower band electrons. This in turn can be very sensitive to the details of the spectrum of electron excitations specific for each metal. Indeed, it has been shown for Au that in the range of electron temperatures typically realized in fs laser material processing applications, thermal excitation of d-band electrons—located around 2 eV below the Fermi level—can lead to a significant increase (up to an order of magnitude) of the e–p coupling factor. This leads to positive deviations of the electron heat capacity from the commonly used linear dependence on the electron temperature [31, 32]. These deviations clearly have to be regarded for a quantitative description of material response to a strong ultra-fast laser excitation for a given material. However, for the heat capacity of the nanoparticle lattice (C_l)

the room temperature values remain under the same conditions and are still reasonable approximations—as C_l does not change much as the temperature increases. For the case of silver, it is known that the change in C_l upon a lattice temperature increase of 1,500 K is less than 20 % compared with its room temperature value of 3.5×10^6 Jm^{-3} K^{-1} [33].

For the electron heat capacity the commonly used linear relationship $C_e(T_e) = \gamma T_e$, with $\gamma = \pi^2 k_B^2 g(\varepsilon_F)/3$, ($g(\varepsilon_F)$: electron density of states (DOS) at the Fermi level) is no longer valid due to the very high electron temperatures reached in the strong perturbation regime. Instead, C_e should include the full spectrum of the electron DOS by taking the derivative of the total electron energy density with respect to the electron temperature [34], as given by the expression

$$C_e(T_e) = \int\limits_{-\infty}^{\infty} \frac{\partial f(\varepsilon, \mu, T_e)}{\partial T_e} g(\varepsilon)\varepsilon d\varepsilon, \tag{3.4}$$

where $g(\varepsilon)$ is the electron DOS at the energy level ε, μ the chemical potential at T_e, and $f(\varepsilon, \mu, T_e)$ is the respective Fermi distribution function.

Here, the e–p coupling is expected to be no longer a constant (typical G values used for silver range from 3×10^{16} to 3.5×10^{16} Wm^{-3} K^{-1} [1, 35, 36].), instead showing considerable temperature dependence for strong excitation. The reason for this is that high electronic temperatures should trigger thermal excitation of the d-band electrons located below the Fermi level, leading to dramatic changes in the rate of the e–p energy exchange. Therefore, the correct treatment of the G factor in the strong perturbation regime requires the consideration of the full spectrum of the electron DOS (as was done for C_e above). The resulting expression for the temperature dependent e–p coupling factor can be given by [37]

$$G(T_e) = \frac{\pi \hbar k_B \lambda \langle \omega^2 \rangle}{g(\varepsilon_F)} \int\limits_{-\infty}^{\infty} g^2(\varepsilon)\left(-\frac{\partial f}{\partial \varepsilon}\right) d\varepsilon, \tag{3.5}$$

where λ denotes the electron–phonon coupling constant, and the value of $\lambda \langle \omega^2 \rangle$ is 22.5 for silver.

Taking into account the theory given above, it is possible to calculate the dependence of the electron heat capacity and the e–p coupling factor on the electronic temperature [21, 38]. The resulting temperature dependence of the electronic heat capacity C_e and the e–p coupling term G turns out to be important when T_e values of above ~5,000 K are achieved. The enhancement of the e–p coupling at high electron temperatures implies a faster energy transfer from the hot electrons to the lattice. A consequence of this temperature dependent e–p coupling term is that the e–p relaxation times (τ_{e-ph}) will increase with increasing electron temperatures and hence the applied laser pulse energy [21, 28, 38]. Therefore, the slightly different e–p relaxation results presented in the literature could be explained by the temperature dependence of the e–p relaxation.

3.3 Two-Temperature Model for the Strong Excitation Regime

We are now going to solve the coupled heat equations (Eqs. 3.1 and 3.2) for a quantitative modeling of the energy relaxation dynamics occurring after the very strong excitation of silver nanoparticles by femtosecond laser pulses. Figure 3.2a depicts the results of 2TM calculations for the case of a single silver nanoparticle excited by a 150 fs pulse with an intensity of 0.5 TW/cm^2 (above the threshold for permanent nanoparticle shape modification [39]) and a central wavelength of 400 nm ($\hbar\omega = 3.1$ eV), close to the SPR. It is easily seen that upon absorbing the laser pulse energy and already during the pulse, conduction electrons of the nanoparticle gain very high temperatures ($\sim 10^4$ K). Reaching the maximum T_e and decaying within a few picoseconds, the hot electronic system heats the cold silver lattice to a region of temperatures above the melting point of (bulk) silver. The electronic and lattice temperatures meet at a value near 2,000 K at ~ 40 ps after the pulse interaction. This suggests the plausibility of the melting of nanoparticles in such a short time frame. Plech et al., using a time-resolved X-ray scattering technique, observed the melting of gold nanoparticles suspended in water within 100 ps after a strong laser pulse excitation [40, 41]. However, it should be noted here that these calculations do not take into account the energy transfer to the matrix and losses due to possible electron emission processes from the nanoparticle, which are additional cooling mechanisms of the electronic sea. The details of these cooling processes will be considered below.

Figure 3.2b shows the dependence of electronic and lattice temperature maxima on a wide range of applied energy densities. The weak regime (up to T_e values of 5,000 K) shows a rapid increase in electronic temperature owing to the very low electronic heat capacity C_e in this interval. However, these electrons do not heat up

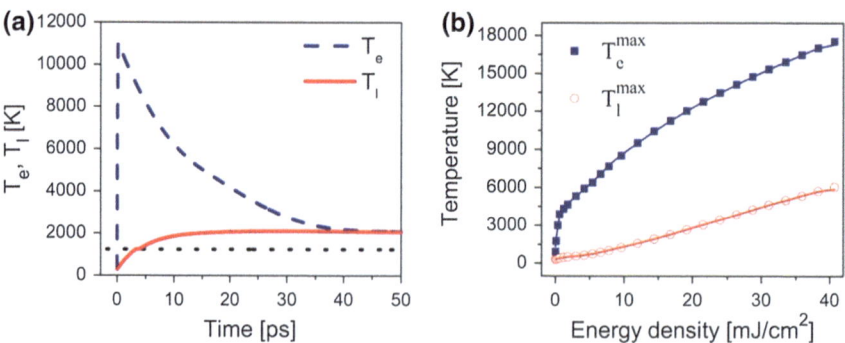

Fig. 3.2 **a** Time evolution of electronic and lattice temperatures of a silver nanoparticle following the absorption of an intense fs laser pulse (around 20 mJ/cm^2 of energy density). The *dotted line* at 1,235 K marks the melting temperature of bulk silver. **b** The dependence of electronic (*blue squares*) and lattice (*red circles*) temperature maxima on a wide range of laser energy densities

the lattice efficiently due to the relatively low e–p coupling factor G. Further increases in the energy density of the pulses cause higher T_e values, but the rise of electronic temperature slows down due to the increasing C_e value. The lattice temperatures are observed to increase with a higher slope in this regime as a result of the increasing efficiency of the G factor. If, for comparison, one uses the standard linear values for C_e (i.e., $C_e(T_e) = \gamma T_e$) at an energy density of 20 mJ/cm^2 (used in the presented 2TM calculations) the 2TM would yield a rise in the electronic temperature to more than 10^5 K, and the resulting T_l values would be much higher than the Ag evaporation temperature.

3.4 Heat Transfer from the Nanoparticle to the Glass Matrix

The above given 2TM describes only the heat transfer between the electrons and the nanoparticle lattice. To get the complete "thermodynamical" picture of the nanoparticle and the surrounding glass system, this 2TM has to be extended by the heat transfer from the nanoparticle to the glass matrix. The excess energy of the nanoparticle is released to the surrounding matrix via phonon couplings across the nanoparticle-glass interface [42, 43]. Therefore, cooling of the nanoparticle (and heating of the glass matrix) can be calculated considering energy flow from the hot particle to the glass through a spherical shell of infinitesimal thickness. Heat transfer from this first heated glass shell is then described by ordinary heat conduction. Because of the large difference in thermal diffusivities of Ag (123 nm^2/ps) and glass (0.5 nm^2/ps), any temperature gradient within the NP can be neglected when calculating the transient temperatures in the surrounding glass (Fig. 3.3).

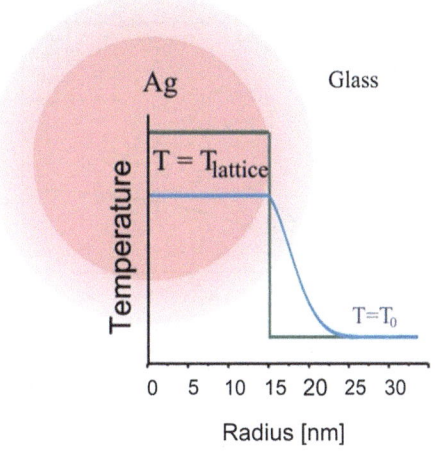

Fig. 3.3 Temperature distribution in a NP-Glass system for different times after irradiation; *green line* ~50 ps, *blue line* a few ns

The temporal and spatial evolution of temperature within the glass can then be calculated by the radial heat equation, where the rate of temperature change ($\partial T(r, t)/\partial t$) is proportional to the curvature of the temperature density ($\partial^2 T(r, t)/\partial r^2$) through the thermal diffusivity (χ) of the glass medium:

$$\frac{\partial T(r,t)}{\partial t} = \frac{\chi}{r} \frac{\partial^2 [rT(r,t)]}{\partial r^2}. \tag{3.6}$$

The timescales for the particle cooling range from tens of picoseconds to nanoseconds, depending on the laser excitation strength, the size of the nanoparticle and the surrounding environment [44].

Figure 3.3 shows the radial temperature distribution in a NP-glass system calculated numerically in the limit of the described 'three-temperature model' (3TM) for two different times after irradiation. After ≈ 50 ps, i.e., when within a spherical NP with radius of 15 nm (red disk in Fig. 3.3) an equilibrated high temperature of $\approx 2,000$ K has been established, the temperature of the surrounding glass matrix is still equal to the room temperature (green line). It takes a few nanoseconds to establish—by energy dissipation into the glass—a heat-affected zone (light magenta circular ring) of the order of 5 nm around the NP (blue line).

More details about the first ten nanoseconds of the time evolution of glass temperatures at different distances away from the nanoparticle are given in Fig. 3.4a. At a distance of 1 nm from the NP surface, the glass is heated up to $T_{max} \approx 1,050$ K within approximately 1 ns after irradiation, then slowly cools down again. With increasing distance of the shells, the maximum temperature decreases. For instance, at a distance of 6 nm a peak value of $T_{max} \approx 500$ K is reached only after ≈ 10 ns. The further evolution of the heat dissipation is shown by some characteristic radial temperature profiles in Fig. 3.4b; here the NP is included, i.e., r = 0 denotes centre of the Ag nanoparticle. At 20 ns the temperatures of the nanoparticle and its nearest shells are around 450–500 K, while the temperature at a distance of 150 nm is still equal to room temperature. After only

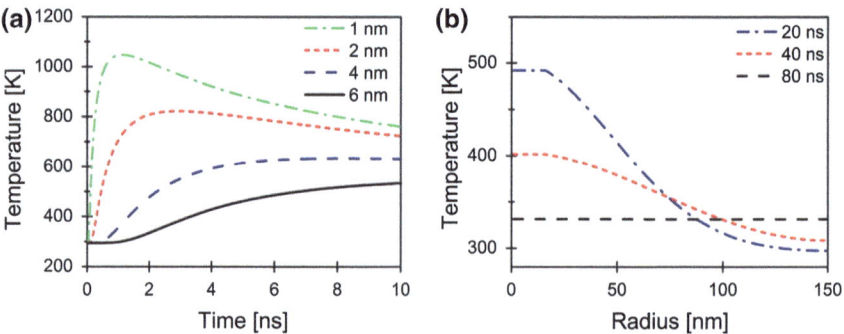

Fig. 3.4 **a** Time evolution of glass temperatures in different shells away from the nanoparticle calculated by 3TM. **b** Temperature distribution for longer times (more then 20 ns) after irradiation calculated by 3TM

80 ns, however, the total energy is nearly homogeneously distributed and the temperature of the layer containing NPs is ≈ 330 K. These calculations have been done for a single metallic nanoparticle of 15 nm radius being surrounded by glass and irradiated by pulses at 400 nm, with intensities of 0.5 TW/cm^2. Remembering that the samples used for fundamental investigations here had an Ag volume filling factor of $\leq 10^{-3}$, which corresponds to an average glass sphere of 150 nm radius around each Ag NP, it is reasonable to regard this model as a well-suited description for our case.

Summarizing the above results one can conclude the following: (1) In the first few ns after the laser pulse the temperature of the NPs is above 1,000 K, and the matrix temperature in the nearest shell up to a distance of 3 nm from the NP can reach or exceed the glass transition temperature [45]; this will cause softening of the glass, which is needed for NP shape transformation [39]. (2) After ≈ 80 ns the system has come to a steady state within the focal volume; from then on heat transfer into the rest of the sample has to be taken into account. It should be mentioned here that this model neglects any glass heating by laser-driven electron and ion emission (which can take place during strong excitation). However, such contributions will only be present within the first few picoseconds after the laser pulse, and will only affect a shell of a few nanometers around the NP [44]. Thus, due to energy conservation the temperature evolution on the timescale of several nanoseconds or slower discussed here should not be affected by this simplification.

3.5 Photoemission from Nanoparticles Incorporated in Dielectric Media

In previous sections the thermo-physical processes due to the interaction of ultra-short laser pulses with nanocomposites were discussed. Now we will proceed to investigate the possible electro-physical processes such as photoemission of electrons and ions from the nanoparticles, which can take place in the strong excitation regime. We will not discuss here the general processes of non-linear ionization that are observed in any kind of material at sufficiently high intensities [46]. Instead, we will concentrate on the following question: how and whether the electric field enhancement in the vicinity of metallic nanoparticles in combination with the strong laser field and the achieved very high temperatures cause the electrons and ions to be released into the glass matrix. Since these processes can have a strong influence on the energy (re-) distribution among the nanoparticles and surrounding dielectric, they are very important for our final understanding of the laser-induced shape transformation of Ag nanoparticles.

In the past, the SP-assisted photoelectron emission from supported Ag nano-particles has been extensively studied upon excitation with intense ultra-short laser pulses [9, 47–53]. The electron work function of the silver clusters—defined as the energy gap between the Fermi level and the energy of the free electron in vacuum—is about 4.3 eV [49, 52, 54]. Moreover, it was demonstrated that excitation near to the SP resonance extremely enhances the two-photon photoemission yield from the Ag nanoparticles [9].

In composite glass containing metal nanoparticles, the probability of SP assisted photoemission can be strongly affected by the structure of the electron energy manifold in the host matrix. In turn, an energy level scheme of the soda-lime glass with embedded Ag nanoparticles can be represented as a dielectric-metal junction (Fig. 3.5). The valence band maximum of the glass lies some 10.6 eV below the vacuum level. The lowest energy level of the conduction band in the glass is placed at some 1.6–1.7 eV below the energy of the free electron in vacuum. Thus, the energy gap between the Fermi level in the silver inclusion (4.3 eV) and conduction band in the glass is about 2.7 eV and consequently any radiation with photon energy >2.7 eV could evoke a tunnel transition of electrons from the silver inclusion into the conduction band of the surrounding glass matrix, even by single photon absorption.

Excitation of the Ag nanoparticles near to the SP resonance (∼3 eV) by a fs laser at 400 nm (3.1 eV) leads to a non-thermal distribution of the electrons in the conduction band of the metal (Fig. 3.5, red dotted line). Since the maximal electron energy in this case exceeds the bottom of the conduction band of the matrix by 0.4 eV, electron injection into the conduction zone of the glass could be possible. At the same time, upon two-photon plasma excitation, the electrons can

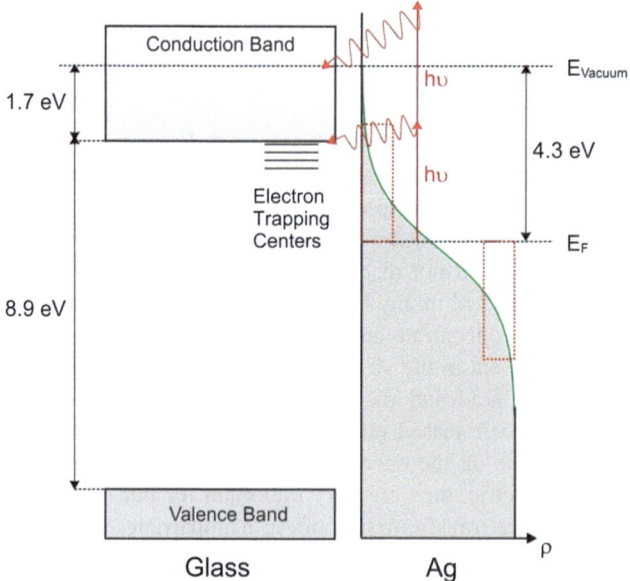

Fig. 3.5 Energy level scheme of the electrons in a composite glass containing silver inclusions. The *red dotted line* indicates a non-thermal distribution of the electrons in the Ag nanoparticle caused by excitation of SP resonance. *Green line* distribution of the electrons after thermalization

overcome the ionization energy level and, unimpeded, penetrate into the glass matrix. In turn, the injection of electrons from metallic inclusions into the conduction band of the surrounding matrix is obviously the origin of an increased conductivity of the composite glass containing Ag nanoparticles upon fs laser irradiation near to the SP resonance [55].

Electrons being emitted during the laser pulse interaction will be driven by a strong, oscillating electric field and therefore generate an anisotropic distribution of emission directions, obviously given by the electric field oscillations (polarization) of the laser pulse. The anticipated 100 fs pulses at $\lambda = 400$ nm correspond to 75 full oscillation cycles with mostly very high amplitudes. A simple estimation shows that a conduction band electron of the nanoparticle can gain a linear acceleration of around 10^{20} m/s^2 upon encountering a linearly polarized pulse of 0.3 TW/cm^2 intensity (corresponding to an electric field amplitude of 10^8 V/m) within the half plasmon period. This is indeed a very large electric field amplitude on the nanoparticle. In the absence of any damping, the above acceleration can push the electron approximately 0.1 nm away from the nanoparticle surface. However, in the case of SP excitation, the oscillation amplitudes of the surface plasmon waves can overcome the excitation amplitude by a few orders of magnitude [56–58]. This means a strong enhancement of the local electromagnetic field in the vicinity of the nanoparticle.

By excitation with polarized light, the E-field enhancement (Fig. 3.6) occurs at special points on the surface. Namely, in the case of spherical nanoparticles, the field is enhanced at the poles of the nanoparticle (Fig. 3.6a), depending on the polarization direction of the exciting light. However, in the case of non-spherical particles (e.g., Fig. 3.6b), it is induced mostly at the tips and corners of the particles [54, 56, 59]. It should be also mentioned here that the local E-field enhancement (EFE) depends on the wavelength (Fig. 3.7); and as it can be seen (Fig. 3.7a), for spherical NPs of radius 15 nm the EFE increases very rapidly to its maximum at the resonance wavelength and then again decreases, whereas for

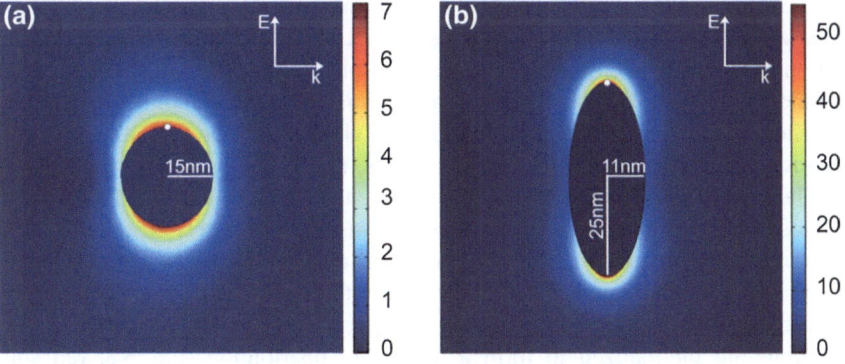

Fig. 3.6 E-field contours for **a** *spherical* (radius 15 nm), and **b** *prolate spheroid* Ag NPs in a glass matrix. Labeled white points illustrate locations for Fig. 3.7

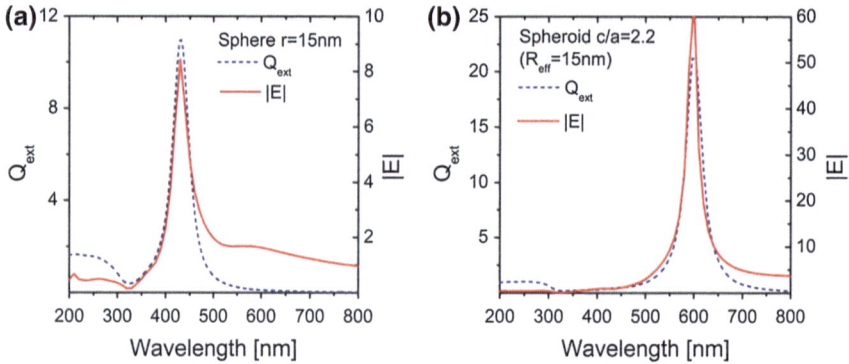

Fig. 3.7 Extinction and electric field enhancement factor along the polarization direction (at the points shown in Fig. 3.6) versus wavelength for **a** *spherical* NPs with $r = 15$ nm, and **b** *prolate spheroids* having a major axis of 25 nm and an aspect ratio of 2.2:1

longer wavelengths its value remains higher than that in the UV spectral region. In the case of non-spherical (spheroidal) particles (Fig. 3.7b) the full spectrum of the electric field enhancement factor (as well as the extinction spectrum) is shifted to the longer wavelength region, and the behavior for the longer and shorter wavelengths stays the same.

As a result of the enhanced electric field at the particle-glass interface, the conduction band electron (discussed above) can move away from the nanoparticle surface by up to a few nm. Electrons driven far away from the nanoparticle leave the region of the strongest field enhancement, and thus will experience a weaker backward force due to the reversed field of the next half plasmon period, and may finally be trapped in the glass matrix. These numbers make plausible that under the specified conditions, there is a non-negligible probability for emission of even 'cold' electrons.

An increase of the electric field in the vicinity of the nanoparticle could strongly suppress energy levels on the metal-dielectric junction and induce effective electron carrier flow from the surface of the nanoparticle—parallel to the laser polarization. The anisotropy in this case is determined by the anomalous distribution of the local electric field over the nanosphere (Fig. 3.6). On the other hand, the electric field in the vicinity of the metal cluster could overcome a breakdown threshold of the glass resulting in high-density electron plasma formation and even ablation of the glass matrix at the poles of the nanosphere.

The thermalization of the electrons with a characteristic time of a few hundred femtoseconds obviously restricts the photoemission processes. However, in the case of a strong excitation, the energy of some electrons could be high enough to jump into the conduction band of the glass [60]. As it was shown above, the maximal electronic temperature after e–e scattering can be higher than 10^4 K. The electrons are thermalized to form a hot Fermi distribution (Fig. 3.8, red solid curve). As can be seen, the high-energy tail of the high T_e Fermi distribution exceeds the energy needed for nanoparticle electrons to penetrate into the glass

Fig. 3.8 Changes in the Fermi distribution of the electronic system following an ultra-short laser pulse irradiation at an energy of 3.1 eV, which excites electrons below the Fermi level to high energies (represented by the *arrows*). The resulting hot electronic distribution is shown with the *solid curve*

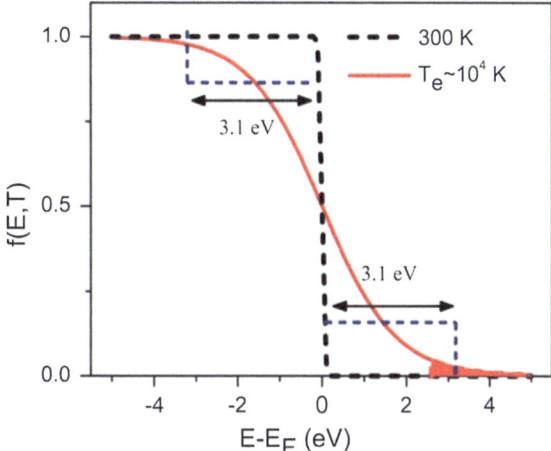

conduction band (red area, >2.6 eV). As long as the pulse is still present, these electrons can be driven by the electric field of the laser along the direction of its polarization. After this time, i.e., even when the pulse is gone, there is still a probability of 'hot' electron emission. However, as the directionality of the pulse is long gone, the thermal emission of the electrons is isotropic.

As a result, irradiation of the silver nanoparticles embedded in glass by strong laser pulses can lead to two different types of electron emission processes, which could be classified as 'pulse-enhanced' or 'purely thermal'. The first one is accordingly anisotropic, the second one isotropic. The isotropic, purely thermal, electron emissions start after the pulse has gone and continue to happen as long as the electrons possess high temperatures (for a few ps). The pulse-enhanced electron emission processes, on the other hand, comprise a 'direct' and a 'pulse-enhanced thermal' electron emission component. The direct electron emission processes are the fastest that happen within the first plasmon oscillation periods. The second component of the pulse-enhanced electron emissions is thermal in nature, owing to the increased electron temperatures along the plasmon oscillation directions. Therefore, in the end, when the emitted electrons will be trapped in the conduction band of the glass, the pulse-enhanced ionization will lead to a non-homogenous electron concentration along the poles of the nanoparticle (along the laser polarization), while the purely thermal electron emission homogeneously spreads the electrons in the nanoparticle's surrounding. It will be shown below that this anisotropic ionization is one of the key processes in the laser-induced nanoparticle shape transformation.

The ionization process leaves the nanoparticles positively charged (due to the emitted electrons), and due to e–p scattering their temperatures rise within a few ps to values of more than 1,000 K (see Sect. 3.3), making the NPs unstable electrically and thermally. Thus, it is obvious that after a few picoseconds the electric potential and thermal energy can overcome the binding energy of the Ag ions, which are being emitted into the surrounding glass matrix [10, 61, 62]. By way of an experimental luminescence study, Podlipensky et al. [62] have proven the presence of Ag ions in the glass matrix emitted upon femtosecond laser irradiation. Using transmission electron microscopy (TEM) after fs laser irradiation small Ag aggregates around the shape-transformed nanoparticle [63, 64] were later observed indicating that this ion emission leads to a partial dissolution of the nanoparticles.

The physical concept behind these ion emission processes is mainly the so-called Coulomb explosion [61], which is a direct consequence of the nanoparticle charging. The repulsive Coulomb forces among the accumulated charges lead to the dissolution (destruction) of the nanoparticle. Even extreme cases of nanoparticle dissolution mechanisms were observed for nanoparticles in an aqueous medium [8], where not only the ions but also some small fragments could leave the nanoparticle because of the soft surrounding. Nevertheless, independent of the way it happens, the total volume of the nanoparticle is reduced over time due to material ejections. In the case of a glass matrix, the isotropically emitted ions can meet the already trapped electrons (a result of the NPs ionization) and recombine with them to form atoms or small silver clusters; the latter can be seen in TEM images [63, 64]. It is obvious that this process of ion ejection will also lead to changes in the energy relaxation (temperature distribution) of the NP-glass system. Some part of the energy will be taken from the nanoparticle and, via kinetic energy of the ions, be transferred to the glass when the ions are trapped there. This will cause much faster heating (compared to merely heat conduction as discussed above) in the first few nanometer shells around a NP.

All the processes discussed so far for the strong excitation regime are in some respect relevant to the fs laser-induced NP shape transformation. We will now develop a self-consistent model for this phenomenon, using all of the above ingredients, which are the results of a number of experimental studies and in-depth analysis of the obtained results.

3.6 Mechanism of the Shape Transformation of Spherical Ag Nanoparticles in Soda-Lime Glass upon fs Laser Irradiation

In order to arrive at a reliable and self-consistent model for the SP-assisted shape modifications of spherical Ag nanoparticles embedded in soda-lime glass upon excitation with intense ultra-short laser pulses, we start with a brief summary of instructive experimental results. The following experimental facts can be stated:

- The process has a threshold [39], i.e., laser-assisted modifications occur only when the laser pulse intensity is sufficiently high (≥ 0.2 TW/cm^2, for excitation at 400 nm).
- Only intensity defines the principal shape of the transformed particles: below ≈ 2 TW/cm^2, the NPs have the shape of prolate spheroids, above this "second threshold" oblate spheroid shapes. The number of laser pulses fired per spot

plays mostly an accumulative role, in particular seen as an increase in the aspect ratio of the prolate NPs with increasing number of pulses [39].

- The anisotropy of shape modifications is strongly correlated with the laser polarization [64, 65]. This indicates that the processes defining the nanoparticle shape are occurring already during the presence of the laser field; so obviously the directed (pulse-enhanced) electron emission from the nanoparticle is an important ingredient of the mechanism.
- Applying laser pulses with very high peak intensity or number of pulses leads to the bleaching and finally disappearance of the extinction SP band [39]; this indicates that partial destruction or dissolution of the silver NPs is involved in the modification mechanism.
- Modification of the NP shapes stops when the wavelength of irradiation is lying in the blue wing of the SPR [66]. However, subsequent irradiation by laser pulses at longer wavelength (or simultaneous two-wavelength irradiation) causes further shape transformation. Therefore, one may presume that the electric field enhancement determines the directional ionization and plays a key role in the mechanism of NP shape transformation.
- Preheating of the sample to up to 150–200 °C frustrates controlled shape transformation; instead total dissolution of the Ag nanoparticles is observed upon laser irradiation [67]. It is anticipated that this effect is due to the increased mobility of silver cations at higher temperature, which prevents the formation of a cation shell in the close vicinity of the NP; in other words, the positively charged shell of silver cations seems to be crucial for the shape modifications.

These conclusions drawn from various experiments show that fs laser-assisted modifications of Ag nanoparticles incorporated in glass are accomplished by a rather complex mechanism. We will therefore first give a general discussion of the anticipated sequence of processes after intense laser excitation, and then specify different cases, i.e., below or above the second threshold, and linear or circular polarization.

The two central effects initiated by the incoming laser pulse are photoionization of the silver nanoparticle and strong heating of the NP and its surroundings. Starting with the latter, we recall that immediately after excitation, the SP relaxes rapidly (within several hundreds of femtoseconds [1, 5]) into a quasi-equilibrated hot electron system via e–e scattering. Next, the hot electrons cool down by sharing their energy with the nanoparticle lattice via e–p couplings [1], thereby heating up the particle. The estimations of maximal electronic and lattice temperature, based on the 2TM (Sect. 3.3), give values in the range of 10^4 K for the electron system and a lattice temperature above the melting temperature for bulk silver. Although the real temperature of the nanoparticle is expected to be much lower (because of nanoparticle energy losses caused by electron and ion emission, see Sect. 3.5), one can conclude that in the course of dissipation of the absorbed laser energy the nanoparticles, and as a result their immediate surroundings, experience a strong transient 'temperature' increase, which is at least connected with the strongly enhanced local mobility of electrons, ions and atoms.

Parallel to heating, photoionization of the silver nanoparticles will occur during the exposure to intense fs laser pulses. The physical idea is that the SPR enhances the electric field of the laser pulse close to an Ag particle by a few orders of magnitude, with the highest fields located at the poles (with respect to the laser polarization) of the nanospheres [56]. This can lead to an intense directed emission of (already hot) electrons from the particle surface [9, 55], preferentially parallel to the laser polarization. But also an isotropic, thermal electron emission has to be regarded in the time frame of the e–p system thermalization. The anisotropy of the direct, laser-driven electron ejection is thought to provide the preferential direction for the following particle shape transformation. Conceivably, the high electric field in the vicinity of the metal nanoparticle can even exceed the breakdown threshold of the glass resulting in high-density electron plasma formation and even ablation of the glass matrix at the poles of the nanosphere. Regardless of whether this happens, the free electrons in the matrix will lead to the formation of colour centres (trapped electrons) in the surroundings of the Ag nanoparticle [62], which also play an important role in particle shape modifications. The free electrons, as well as the colour centres, have strong absorption at the SP resonance [68], which might result in resonant coupling of the SP oscillations to the matrix [69]. Finally, the ionized positively charged core of the Ag nanoparticles becomes unstable and the Coulomb forces will eject silver cations, which form a cationic shell in the vicinity of the nanoparticle [62]. Clearly the radius of such a cationic shell is determined by the diffusion length of the silver cations and thus strongly depends on the local temperature.

All these effects are transient phenomena, being controlled either directly by the electric field of the laser pulse or indirectly by the induced temperature rise. Thus, the pulse intensity is doubtlessly the decisive parameter for the shape transformation of the metal nanoparticles and it is obvious to assume that the prevalence of individual processes, due to their different intensity dependence, leads to the characteristic intensity regions yielding prolate or oblate shapes. Our investigations strongly suggest the following picture.

Using linear polarization in the low intensity region (i.e., between $I_1 \approx 0.2$ TW/cm^2 and $I_2 \approx 2$ TW/cm^2) [39], SP field enhancement stimulates the fastest process, field-driven electron emission from the surface of the metal particles (Fig. 3.9a). The emission process happens within a few fs [10], i.e., without delay against the laser pulse. The ejected electrons will then be trapped in the matrix forming colour centres close to the poles of the sphere. The ionized nanoparticles are likely to emit Ag ions in statistical directions, in particular after a few picoseconds when electron thermalization and heat transfer to the silver lattice is finished. The resulting positively charged shell of silver cations [62] around the particle meets trapped electrons that are mostly concentrated at the poles (Fig. 3.9b). After some picoseconds they can recombine to Ag atoms (Fig. 3.9c), which finally diffuse back to the nanoparticle and precipitate mainly at the poles. Silver atoms that are situated relatively far away from the main nanoparticle can locally precipitate with each other forming very small clusters (halo). In the multi-shot pulsed laser mode, remaining silver ions may also act as trapping centres for

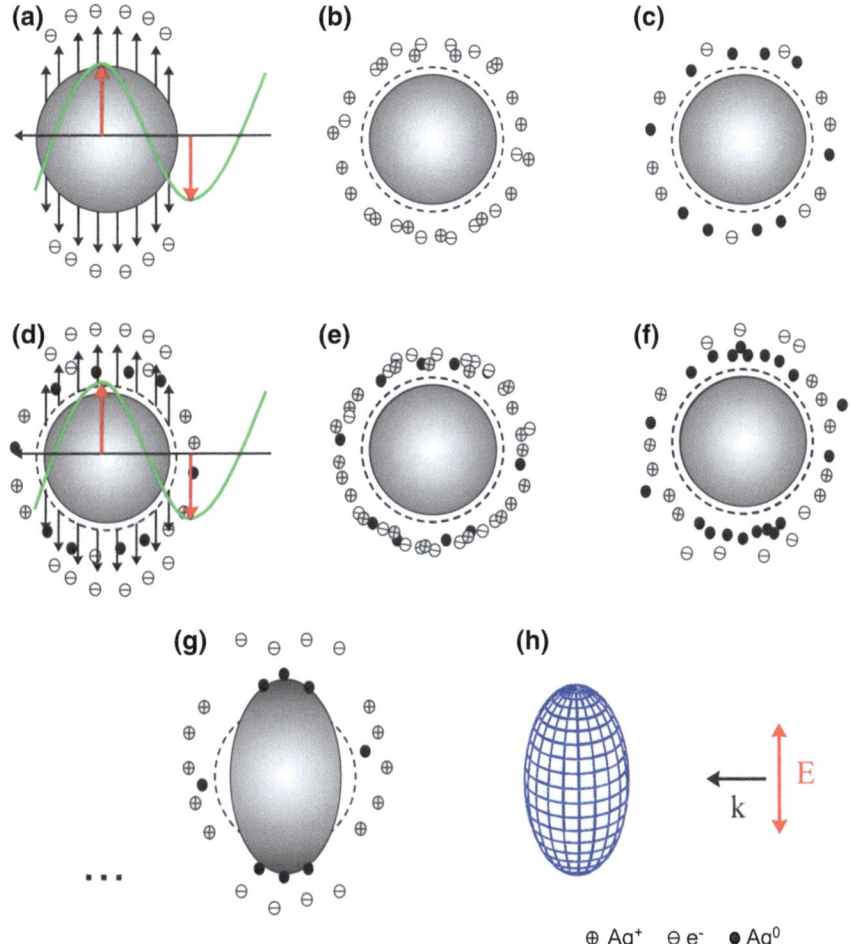

Fig. 3.9 Laser-assisted shape transformation of the metal nanospheres in the case of irradiation by *linearly polarized* laser pulses in low intensity, multi-shot mode. See text for explanation

the electrons being emitted by the next laser pulses (Fig. 3.9d–f). Possibly also the fact that electrons are favorably being trapped close to the poles—while ions which are mostly concentrated around equator (because the purely thermal emission of electrons leads to fewer electrons being available for ion annihilation there)—may cause local electric field distributions which influence the directional properties of the electron and ion emission for the subsequent laser pulses.

All these processes lead to a step-by-step growth of the Ag particles along the laser polarization, explaining the eventually observed prolate spheroidal shape [64] (Fig. 3.9g, h). Especially, during the growing process most of the very small silver clusters precipitated above the poles (defined by the laser polarization) become closer to the main nanoparticle and can be incorporated into it again, while the clusters situated around equator contribute only to the halo formation [64]. At the same time, it is obvious that in the case of circular polarization the rotating electric field should lead to precipitation around the equator of a NP, rather than at the (no longer defined) poles, resulting in an oblate spheroidal shape [65] with the halo now forming in a perpendicular direction (i.e., the direction of propagation). Figure 3.10 illustrates this case in analogy to Fig. 3.9.

With increasing peak pulse intensity we expect a higher temperature and thus a larger radius of the cationic shell. In this case, the farthest clusters located even in the direction of laser polarization will not diffuse back to the main nanoparticle, and in consequence a larger halo region is observed [39]. It should be mentioned that all the processes discussed here require the presence of a rigid, ionic matrix. This may explain why thus far the laser-induced transformation of metal nano-particles has only been observed in glass nanocomposites.

In the high intensity range (above 2 TW/cm^2) [39] and using linear polarization, we suggest that in addition to the processes already discussed, the very high local electric field at the poles of the sphere (along the laser polarization) can accelerate the free electrons so strongly that they induce a high density plasma by avalanche ionization of the glass (Fig. 3.11a). The following plasma relaxation transfers energy from the electrons to the lattice (e–p interaction) on a timescale much faster than the thermal diffusion time. This can ultimately result in ablation of the material at the interface between the glass matrix and the metal inclusion thereby leading to the partial destruction of the nanoparticle at its poles (Fig. 3.11b), or direct emission of the plasma components further away into the matrix. In any case, the process

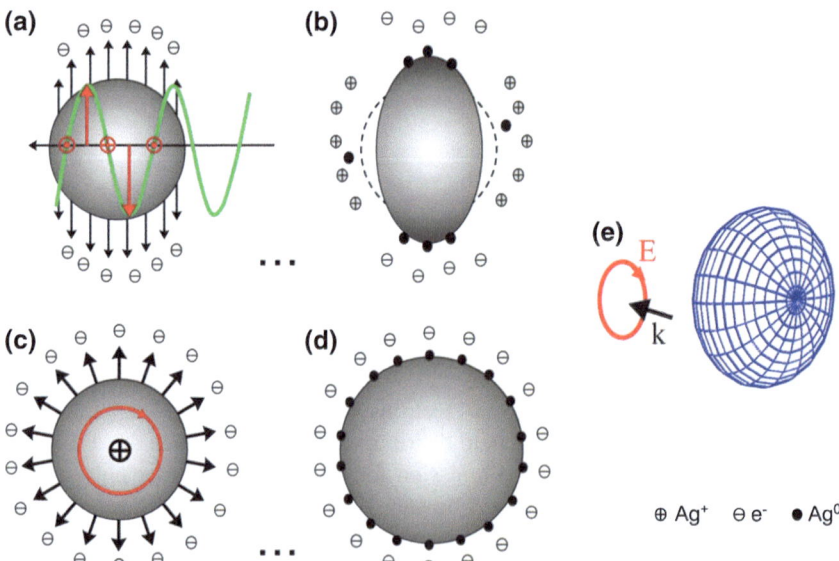

Fig. 3.10 Laser-assisted shape transformation of metal nanospheres in the case of irradiation by *circularly polarized* laser pulses

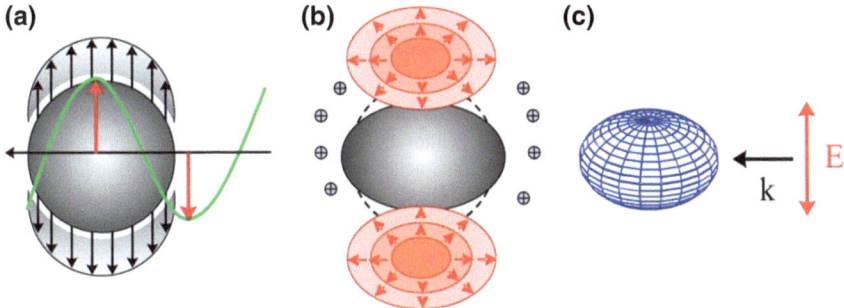

Fig. 3.11 Laser-assisted shape transformation of metal nanospheres in the case of irradiation by *linearly polarized* laser pulses with high intensities

produces oblate rather than prolate particle shapes. It seems reasonable to anticipate that the characteristic intensity ($I_2 \approx 2$ TW/cm^2) marks the balance between (1) the processes leading to particle growth along the laser polarization and (2) the beginning of plasma formation at the particle/matrix interface counteracting this growth.

We emphasise that for any type of shape transformation of spherical Ag nanoparticles embedded in soda-lime glass induced by fs laser pulses, the main point of the proposed scenario is the SP assisted (field-driven) photoelectron emission from the metal surface, which defines the particle symmetry. The next steps of the mechanism 'only' decide—via the applied laser pulse intensity and the resulting local heating—whether the nanoparticles should develop their final shape in an accumulative process (multi-shot mode) or if, at too high an intensity, a few or even one laser pulse is sufficient to destroy the particle completely.

Having elaborated the physical model for NP shape transformation we will now, in the following sections, describe in some detail several irradiation and sample parameters in terms of their effect on the degree of nanoparticle deformation. This discussion will not only confirm the proposed mechanism but also pave the way for this innovative technology to be put to good practical use.

References

1. Bigot, J.Y., Halte, V., Merle, J.C., Daunois, A.: Electron dynamics in metallic nanoparticles. Chem. Phys. **251**, 181–203 (2000)
2. Link, S., El-Sayed, M.A.: Shape and size dependence of radiative, non-Rradiative and photothermal properties of gold nanocrystals. Int. Rev. Phys. Chem. **19**, 409–453 (2000)
3. Arbouet, A., Voisin, C., Christofilos, D., Langot, P., Del Fatti, N., Vallee, F., Lerme, J., Celep, G., Cottancin, E., Gaudry, M., Pellarin, M., Broyer, M., Maillard, M., Pileni, M.P., Treguer, M.: Electron-Phonon Scattering in Metal Clusters. Phys. Rev. Lett. **90**, 177401 (2003)

4. Voisin, C., Christofilos, D., Loukakos, P.A., Del Fatti, N., Vallee, F., Lerme, J., Gaudry, M., Cottancin, E., Pellarin, M., Broyer, M.: Ultrafast electron–electron scattering and energy exchanges in noble-metal nanoparticles. Phys. Rev. B **69**, 195416 (2004)
5. Del Fatti, N., Voisin, C., Achermann, M., Tzortzakis, S., Christofilos, D., Vallee, F.: Nonequilibrium electron dynamics in noble metals. Phys. Rev. B **61**, 16956 (2000)
6. Del Fatti, N., Arbouet, A., Vallee, F.: Femtosecond optical investigation of electron–lattice interactions in an ensemble and a single metal nanoparticle. Appl. Phys. B **84**, 175–181 (2006)
7. Stietz, F.: Laser manipulation of the size and shape of supported nanoparticles. Appl. Phys. A **72**, 381–394 (2001)
8. Kamat, P., Flumiani, M., Hartland, G.V.: Picosecond Dynamics of Silver Nanoclusters. Photoejection of Electrons and Fragmentation. J. Phys. Chem. B **102**, 3123–3128 (1998)
9. Pfeiffer, W., Kennerknecht, C., Merschdorf, M.: Electron dynamics in supported metal nanoparticles: Relaxation and charge transfer studied by time-resolved photoemission. Appl. Phys. A **78**, 1011–1028 (2004)
10. Calvayrac, F., Reinhard, P.G., Suraud, E., Ullrich, C.A.: Nonlinear electron dynamics in metal clusters. Phys. Rep. **337**, 493–578 (2000)
11. Fann, W.S., Storz, R., Tom, H.W.K., Bokor, J.: Direct measurement of nonequilibrium electron-energy distributions in subpicosecond laser-heated gold films. Phys. Rev. Lett. **68**, 2834–2837 (1992)
12. Fann, W.S., Storz, R., Tom, H.W.K., Bokor, J.: Electron thermalization in gold. Phys. Rev. B **46**, 13592–13595 (1992)
13. Perner, M., Bost, P., Lemmer, U., von Plessen, G., Feldmann, J., Becker, U., Mennig, M., Schmitt, M., Schmidt, H.: Optically Induced Damping of the Surface Plasmon Resonance in Gold Colloids. Phys. Rev. Lett. **78**, 2192–2195 (1997)
14. Aeschlimann, M., Bauer, M., Pawlik, S.: Competing nonradiative channels for hot electron induced surface photochemistry. Chem. Phys. **205**, 127–141 (1996)
15. Cao, J., Gao, Y., Elsayed-Ali, H.E., Miller, R.J.D., Mantell, D.A.: Femtosecond photoemission study of ultrafast electron dynamics in single-crystal Au (111) films. Phys. Rev. B **58**, 10948–10952 (1998)
16. Knoesel, E., Hotzel, A., Wolf, M.: Ultrafast dynamics of hot electrons and holes in copper: Excitation, energy relaxation, and transport effects. Phys. Rev. B **57**, 12812–12824 (1997)
17. Ogawa, S., Nagano, H., Petek, H.: Hot-electron dynamics at Cu(100), Cu(110), and Cu(111) surfaces: mComparison of experiment with Fermi-liquid theory. Phys. Rev. B **55**, 10869–10877 (1997)
18. Voisin, C., Christofilos, D., Del Fatti, N., Vallee, F., Prevel, B., Cottancin, E., Lerme, J., Pellarin, M., Broyer, M.: Size-Dependent Electron–Electron Interactions in Metal Nanoparticles. Phys. Rev. Lett. **85**, 2200–2203 (2000)
19. Voisin, C., Del Fatti, N., Christofilos, D., Vallee, F.: Ultrafast Electron Dynamics and Optical Nonlinearities in Metal Nanoparticles. J. Phys. Chem. B **105**, 2264–2280 (2001)
20. Hodak, J., Martini, I., Hartland, G.V.: Ultrafast study of electron-phonon coupling in colloidal gold particles. Chem. Phys. Lett. **284**, 135–141 (1998)
21. Ahmadi, T.S., Logunov, S.L., El-Sayed, M.A.: Picosecond Dynamics of Colloidal Gold Nanoparticles. J. Phys. Chem. **100**, 8053–8056 (1996)
22. Roberti, T.W., Smith, B.A., Zhang, J.Z.: Ultrafast electron dynamics at the liquid–metal interface: Femtosecond studies using surface plasmons in aqueous silver colloid. J. Chem. Phys. **102**, 3860–3866 (1995)
23. Haus, J.W., Kalyaniwalla, N., Inguva, R., Bloemer, M., Bowden, C.M.: Nonlinear-optical properties of conductive spheroidal particle composites. J. Opt. Soc. Am. B **6**, 797–807 (1989)
24. Uchida, K., Kaneko, S., Omi, S., Hata, C., Tanji, H., Asahara, Y., Ikushima, A.J., Tokizaki, T., Nakamara, A.: Optical nonlinearities of a high concentration of small metal particles dispersed in glass: Copper and silver particles. J. Opt. Soc. Am. B **11**, 1236–1243 (1994)

25. Stella, A., Nisoli, M., De Silvestri, S., Svelto, O., Lanzani, G., Cheyssac, O., Kofman, R.: Size effects in the ultrafast electronic dynamics of metallic tin nanoparticles. Phys. Rev. B **53**, 15497–15500 (1996)

26. Kreibig, U., Bour, G., Hilger, A., Gartz, M.: Optical Properties of Cluster–Matter: Influences of Interfaces. Phys. Status Solidi A **175**, 351–366 (1999)

27. Bigot, J.Y., Merle, J.C., Cregut, O., Daunois, A.: Electron Dynamics in Copper Metallic Nanoparticles Probed with Femtosecond Optical Pulses. Phys. Rev. Lett. **75**, 4702–4705 (1995)

28. Hodak, J.H., Martini, I., Hartland, G.V.: Spectroscopy and Dynamics of Nanometer-Sized Noble Metal Particles. J. Phys. Chem. B **102**, 6958–6967 (1998)

29. Anisimov, S.I., Kapeliovich, B.L., Perelman, T.L.: Electron emission from metal surface exposed to ultrashort laser pulse. Zh. Eksp. Teor. Fiz. **66**, 776 (1974)

30. Muskens, O.L., Del Fatti, N., Vallee, F.: Femtosecond Response of a Single Metal Nanoparticle. Nano Lett. **6**, 552–556 (2006)

31. Wang, X.Y., Riffe, D.M., Lee, Y.S., Downer, M.C.: Time-resolved electron-temperature measurement in a highly excited gold target using femtosecond thermionic emission. Phys. Rev. B **50**, 8016–8019 (1994)

32. Lin, Z., Zhigilei, L.V.: Thermal excitation of d band electrons in Au: implications for laser-induced phase transformations. Proc. SPIE **6261**, 62610U (2006)

33. Hultgren, R., Desai, P.D., Hawkins, D.T., Gleiser, M., Kelley, K.K., Wagman, D.D.: Selected Values of the Thermodynamic Properties of the Elements. ASM, Metals Park (1973)

34. Ashcroft, N.W., Mermin, N.D.: Solid State Phys. HCP, Orlando (1976)

35. Hamanaka, Y., Hayashi, N., Nakamura, A., Omi, S.: Ultrafast relaxation dynamics of electrons in silver nanocrystals embedded in glass. J. Lumin. **76**, 221–225 (1998)

36. Lysenko, S., Jimenez, J., Zhang, G., Liu, H.: Nonlinear optical dynamics of glass-embedded silver nanoparticles. J. Electron Mater. **35**, 1715–1721 (2006)

37. Lin, Z., Zhigilei, L.V., Celli, V.: Electron-phonon coupling and electron heat capacity of metals under conditions of strong electron-phonon nonequilibrium. Phys. Rev. B **77**, 075133 (2008)

38. Kresse, G., Hafner, Ab initio molecular dynamics for liquid metals. J.: Phys. Rev. B **47**, 558–561 (1993)

39. Stalmashonak, A., Podlipensky, A., Seifert, G., Graener, H.: Intensity-driven, laser induced transformation of Ag nanospheres to anisotropic shapes. Appl. Phys. B **94**, 459–465 (2009)

40. Plech, A., Kotaidis, V., Gresillon, S., Dahmen, C., von Plessen, G.: Laser-induced heating and melting of gold nanoparticles studied by time-resolved x-ray scattering. Phys. Rev. B **70**, 195423 (2004)

41. Plech, A., Kotaidis, V., Lorenc, M., Boneberg, J.: Femtosecond laser near-field ablation from gold nanoparticles. Nat. Phys. **2**, 44–47 (2006)

42. Inouye, H., Tanaka, K., Tanahashi, I., Hirao, K.: Ultrafast dynamics of nonequilibrium electrons in a gold nanoparticle system. Phys. Rev. B **57**, 11334–11340 (1998)

43. Plech, A., Kurbitz, S., Berg, K.J., Graener, H., Berg, G., Gresillon, S., Kaempfe, M., Feldmann, J., Wulff, M., von Plessen, G.A.: Time-resolved X-ray diffraction on laser-excited metal nanoparticles. Europhys. Lett. **61**, 762–768 (2003)

44. Unal, A.A., Stalmashonak, A., Graener, H., Seifert, G.: Time-resolved investigation of laser-induced shape transformation of silver nanoparticles. Phys. Rev. B **80**, 115415 (2009)

45. Podlipensky, A., Abdolvand, A., Seifert, G., Graener, H.: Femtosecond laser assisted production of dichroic 3D structures in composite glass containing Ag nanoparticles. Appl. Phys. A **80**, 1647–1652 (2005)

46. Schaffer, C.B. Interaction of femtosecond laser pulses with transparent materials. PhD thesis, Harvard University. http://mazur-www.harvard.edu/publications.php?function=display&rowid=123 (2001)

47. Busolt, U., Cottancin, E., Rohr, H., Socaciu, L., Leisner, T., Woste, L.: Eur. Phys. J. D **9**, 523–527 (1999)

48. Fierz, M., Siegmann, K., Scharte, M., Aeschlimann, M.: Appl. Phys. B **68**, 415–418 (1999)

49. Ertel, K., Kohl, U., Lehmann, J., Merschdorf, M., Pfeiffer, W., Thon, A., Voll, S., Gerber, G.: Appl. Phys. B **68**, 439–445 (1999)
50. Koller, L., Schumacher, M., Kohn, J., Teuber, S., Tiggesbaumker, J., Meiwes-Broer, K.H.: Phys. Rev. Lett. **82**, 3783–3786 (1999)
51. Lehmann, J., Merschdorf, M., Pfeiffer, W., Thon, A., Voll, S., Gerber, G.: Phys. Rev. Lett. **85**, 2921–2924 (2000)
52. Lehmann, J., Merschdorf, M., Pfeiffer, W., Thon, A., Voll, S., Gerber, G.: J. Chem. Phys. **112**, 5428–5434 (2000)
53. Merschdorf, M., Pfeiffer, W., Thon, A., Voll, S., Gerber, G.: Appl. Phys. A **71**, 547–552 (2000)
54. Kreibig, U., Vollmer, M.: Optical Properties of Metal Clusters, vol. 25. Springer Series in Material ScienceSpringer, Berlin, DE (1995)
55. Akella, A., Honda, T., Liu, A.Y., Hesselink, L.: Two-photon holographic recording in aluminosilicate glass containing silver particles. Opt. Lett. **22**, 967–969 (1997)
56. Kelly, K.L., Coronado, E., Zhao, L.L., Schatz, G.C.: The Optical Properties of Metal Nanoparticles: The Influence of Size, Shape, and Dielectric Environment. J. Phys. Chem. B **107**, 668–677 (2003)
57. Hao, E., Schatz, G.C.: Electromagnetic fields around silver nanoparticles and dimers. J. Chem. Phys. **120**, 357–366 (2004)
58. Hao, E., Schatz, G.C., Hupp, J.T.: Synthesis and Optical Properties of Anisotropic Metal Nanoparticles. J. Fluoresc. **14**, 331–341 (2004)
59. Shalaev, V.M., Kawata, S.: Nanophotonics with surface plasmons. In: Advances in Nano-Optics and Nano-Photonics. Elsevier, UK (2007)
60. Unal, A.A., Stalmashonak, A., Seifert, G., Graener, H.: Ultrafast dynamics of silver anoparticle shape transformation studied by femtosecond pulse-pair irradiation. Phys. Rev. B **79**, 115411 (2009)
61. Doppner, T., Fennel, T., Diederich, T., Tiggesbaumker, J., Meiwes-Broer, K.H.: Controlling the Coulomb Explosion of Silver Clusters by Femtosecond Dual-Pulse Laser Excitation. Phys. Rev. Lett. **94**, 013401 (2005)
62. Podlipensky, A.V., Grebenev, V., Seifert, G., Graener, H.: Ionization and photomodification of Ag nanoparticles in soda-lime glass by 150 fs laser irradiation: a luminescence study. J. Lumin. **109**, 135–142 (2004)
63. Kaempfe, M., Seifert, G., Berg, K.-J., Hofmeister, H., Graener. H.: Polarization dependence of the permanent deformation of silver nanoparticles in glass by ultrashort laser pulses. Eur. Phys. J. D **16**, 237–240 (2001)
64. Stalmashonak, A., Seifert, G., Graener, H.: Optical three-dimensional shape analysis of metallic nanoparticles after laser-induced deformation. Opt. Lett. **32**, 3215–3217 (2007)
65. Stalmashonak, A., Graener, H., Seifert, G.: Transformation of silver nanospheres embedded in glass to nanodisks using circularly polarized femtosecond pulses. Appl. Phys. Lett. **94**, 193111 (2009)
66. Stalmashonak, A., Seifert, G., Graener, H.: Spectral range extension of laser-induced dichroism in composite glass with silver nanoparticles. J. Opt. A: Pure Appl. Opt. **11**, 065001 (2009)
67. Stalmashonak, A., Unal, A.A., Graener, H., Seifert, G.: Effects of Temperature on Laser-Induced Shape Modification of Silver Nanoparticles Embedded in Glass. J. Phys. Chem. C **113**, 12028–12032 (2009)
68. Podlipensky, A.V.: Laser assisted modification of optical and structural properties of composite glass with silver nanoparticles. Ph.D. Thesis, Martin-Luther-Universität Halle-Wittenberg. http://sundoc.bibliothek.uni-halle.de/dissonline/05/05H084/t1.pdf (2005)
69. Melikyan, A., Minassian, H.: On surface plasmon damping in metallic nanoparticles. Appl. Phys. B **78**, 453–455 (2004)

Chapter 4
Effect of Pulse Intensity and Writing Density on Nanoparticle Shape

We start this discussion with results obtained by applying laser pulses of quite different peak intensities and 'writing density'; the latter referring to the question of how many pulses are, on average, hitting one position on the sample. In particular, the following questions will be addressed: (1) is there a single pulse intensity threshold for deformation, or can lower intensity be compensated for by firing more pulses? (2) at which intensity/number of pulses applied does the transition from prolate to oblate shape occur, and which particle shapes are produced there? (3) what happens with the particles going to very high irradiation intensity and/or large number of pulses applied to one spot?

For measurements of the intensity dependences we used a technique of space resolved transmission spectra, which has been described in some length previously [1]. In combination with laser beam profile measurements the space resolved spectra were correlated with local laser pulse intensities. We have produced various dichroic areas on the sample by irradiating different numbers of laser pulses (ranging from 1 to 5,000) onto the same spot. The large number of spectra resulting from the described analysis will be shown here in a parameterized form (see below). Nonetheless, to demonstrate the quality of the spectra and explain the parameterization, a few examples are shown in Fig. 4.1.

Figures 4.1a, c represent the case of multi-shot, whilst Fig. 4.1b, d that of a single-shot irradiation. In general, the original SPR band, peaked at $\lambda = 413$ nm, splits into two polarization dependent bands upon irradiation, but with significant dependence on peak pulse intensity and the number of pulses applied. Figure 4.1a shows multi-shot irradiation (1,000 pulses at 0.6 TW/cm^2), which produces bands on different sides of the original SPR band: for polarization parallel to that of the laser (p-polarized, blue line), the peak position is shifted to longer wavelengths, while for perpendicular polarization (s-polarized, red line) the band is observed at a shorter wavelength.

This can be explained on the nanoscale by prolate silver spheroids with their symmetry axes oriented along the laser polarization (see previous section) (inset in Fig. 4.1a). In the single-shot case (Fig. 4.1b, referring to 3 TW/cm^2), the s–polarized band has a larger red-shift than the p-polarized band. Additionally, both bands

A. Stalmashonak et al., *Ultra-Short Pulsed Laser Engineered Metal–Glass Nanocomposites*, SpringerBriefs in Physics, DOI: 10.1007/978-3-319-00437-2_4, © The Author(s) 2013

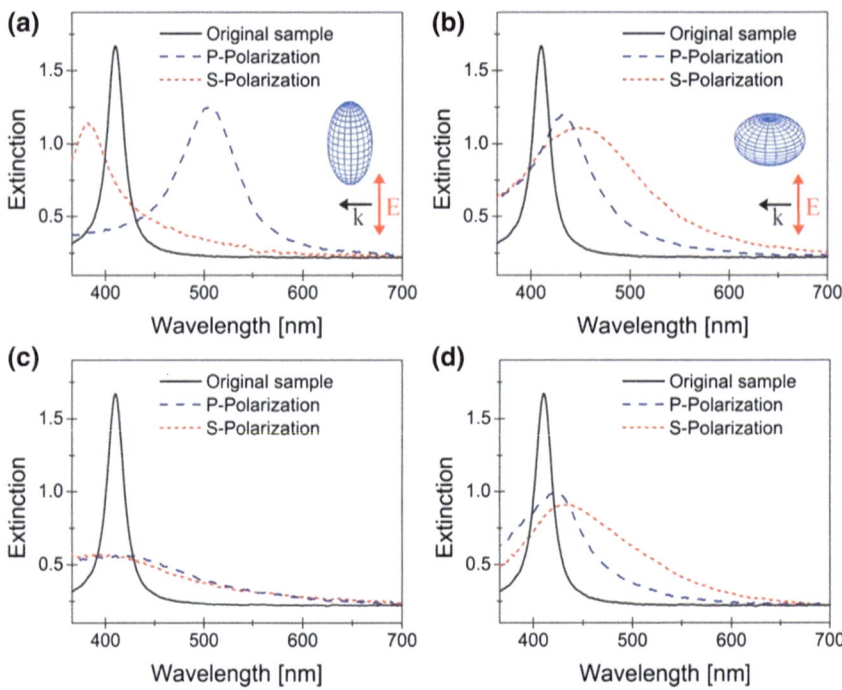

Fig. 4.1 Polarized extinction spectra of original and irradiated samples **a** multi-shot regime (1,000 pulses per spot), peak pulse intensity $I_p = 0.6$ TW/cm^2, **b** single-shot regime, $I_p = 3$ TW/cm^2, **c** multi-shot (5,000 pulses per spot), $I_p = 1.2$ TW/cm^2, **d** single-shot, $I_p = 3.5$ TW/cm^2

are red-shifted in this case. These spectra are due to oblate Ag particles (inset in Fig. 4.1b), once again with their symmetry axes oriented along the laser polarization [2]. At even higher intensities and, in particular, in the multi-shot regime, the spectral shifts are becoming smaller and the band integrals decrease. These effects, which are obviously indicating—at least partial—destruction of the silver nanoparticles, are most clearly seen in Fig. 4.1c (representing 5,000 pulses at 1.2 TW/cm^2), but tentatively also in the single-shot regime (Fig. 4.1d, 3.5 TW/cm^2).

The extreme cases of single- and multi-shot compared in Fig. 4.1 pose the question of how a continuous variation of peak intensity and number of applied pulses affect the resulting spectral parameters such as orientation of dichroism, peak position and integrated extinction of the SPR bands. Figure 4.2 presents a selection of pertinent results in a parametrized form: the peak positions of the two SP bands observed with polarization perpendicular to each other ('p' referring to parallel with respect to the laser polarization) are given as a function of intensity, with an increasing number of applied pulses from (a) to (f). It is seen that generally laser-induced spectral changes start at intensities of 0.2–0.3 TW/cm^2. For single-shot irradiation (Fig. 4.2a), increase of pulse intensity above this threshold leads to a shift of both SP bands towards longer wavelengths. First, in the region of

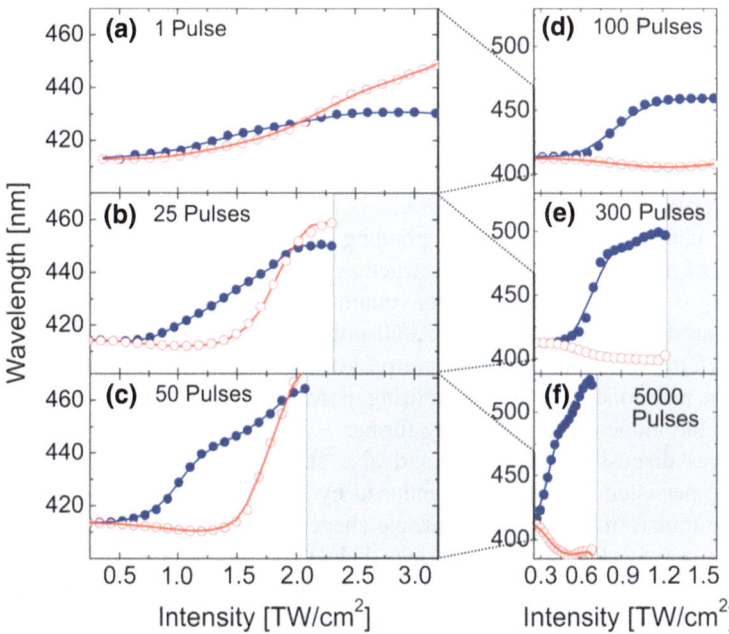

Fig. 4.2 Dependence of the SP maximum in polarized extinction spectra of soda-lime glass with spherical Ag nanoparticles on laser pulse intensity by irradiation at 400 nm. *Red circles* s-polarization, *blue solid circles* p-polarization, *light gray area* region of SP bleaching

~ 0.4 TW/cm^2 one observes a rather weak dichroism, where the p-polarized SP band has the stronger red-shift. At approximately 2 TW/cm^2 the extinction is becoming isotropic again, seen as a crossing of the curves for s- and p-polarization at $\lambda = 427$ nm. Above 2 TW/cm^2 a reversed dichroism is observed, i.e., the s-polarization band is now more red-shifted than the p-band. The maximum spectral gap between p- and s-polarized SP bands (peaks at 430 and 450 nm, respectively) is found at ~ 3.2 TW/cm^2. At still higher intensity, beyond 3.2 TW/cm^2 (not shown on the Fig. 4.2), the SP bands move back toward shorter wavelengths and the integrated band extinction decreases, indicating (partial) destruction of the silver nanoparticles.

Irradiating 25 pulses onto one spot (Fig. 4.2b) we find in general a similar peak pulse intensity dependence of the induced dichroism with two characteristic intensity ranges. There are, however, some important differences compared to the single-shot case: (1) the dichroism (spectral spacing between the polarization dependent bands) is much larger in the low intensity range (below 2 TW/cm^2); (2) between 0.3 and 1.3 TW/cm^2 the s-polarized SP band is blue-shifted relative to the original SP peak at 413 nm; (3) the region of reversed dichroism has shrunk considerably, because from ~ 2.3 TW/cm^2 onwards bleaching of the extinction (particle destruction) starts already. In such cases the analysis of the SP peak central wavelengths was halted (grey regions in Fig. 4.2).

If the number of pulses irradiated onto one spot is further increased one observes that the maximum dichroism grows and is achieved at lower peak pulse intensity (Figs. 4.2c–f; note the scales change from Fig. 4.2c–d). The crossing point of the curves for the p- and s-band, however, remains approximately constant around 2 TW/cm^2, while the onset of particle destruction reduces incrementally in intensity to finally reach $\sim$0.7 TW/cm^2 at 5,000 pulses per spot (Fig. 4.2f). Thus, for 100 or more pulses per spot we can only observe the low intensity region of spectral changes with the corresponding dichroism, because the subsequent increase of intensity leads to the destruction of the nanoparticles and results in the bleaching of the SP bands. The maximum dichroism recognized in our experiments was found in the case of 5,000 pulses, where, at the pulse intensity of 0.65 TW/cm^2, the p- and s-bands are peaked at 525 and 390 nm, respectively. It should be mentioned here that even firing more than 5,000 pulses per spot does not increase the induced dichroism any further.

As was discussed previously, and also shown in Refs. [3–6], and [7], the principal persistent modifications induced by fs laser pulses does not only comprise the transformation of nanoparticle shapes, but also the generation of a surrounding region of small Ag particles ('halo'). While the first effect explains the splitting of the SP band (dichroism), the second one causes, to a first approximation, a modified matrix refractive index that may lead to an isotropic spectral shift of the SP bands [2].

With this additional information, the above presented results on peak intensity and irradiation density (number of pulses per spot) of the fs laser pulses allow us the following general conclusions: independent of the number of pulses applied, there exist two special intensities $I_1 \approx 0.2$ TW/cm^2 and $I_2 \approx 2$ TW/cm^2. For intensities $I < I_1$ there is no spectral change at all, and at $I = I_2$ only a spectral shift of the SP band to longer wavelengths is observed. In the intensity region $I_1 < I < I_2$, dichroism is found with the larger red-shift for the p-polarized SP band, while for $I > I_2$ a reversed dichroism is seen. This indicates that the processes of shape transformation are controlled by the laser pulse intensity (energy density per pulse), while the number of pulses applied mainly accumulates the changes caused by each single pulse. Looking in more detail to the low-intensity region, $I_1 < I < I_2$, first, the dichroism observed there is associated with a transformation of the original silver nanospheres to prolate spheroids with their long axis oriented parallel to the laser polarization [2]. Anticipating volume conservation for the silver, Mie theory predicts for this case a blue- (red-) shift of the SPR for the short (long) particle axis, the spectral spacing between the two bands being correlated to the aspect ratio of the nanoparticle. Consequently the growth of dichroism with increasing number of pulses can be explained by a successive increase of the particle's aspect ratio.

A red-shift of both bands however, as observed for 1 pulse of any intensity or at $I > 1.5$ TW/cm^2 for 25 or 50 pulses, can only be explained by additional modification of the host matrix in the vicinity of the nanoparticle, which was shown in Refs. [2, 7]. Thus, it is obvious to assign the increasing general red-shift for higher pulse intensities to a growing influence of the halo. In the high-intensity region,

$I > I_2$, oblate spheroids with their symmetry axes (short axis) along the laser polarization are produced. Again the fact that both SP bands are red-shifted indicates significant modification of the particle surroundings, because otherwise the short axis should show a blue-shifted SP band.

To get an idea about the nanoscopic modifications in the region around $I_2 \approx 2$ TW/cm^2, and in the onset region for particle destruction (where the SP band extinction starts to decrease again), transmission electron microscopy is quite instructive. It should be mentioned, however, that it is not possible to assign any particular local irradiation intensity to a special TEM image. Figure 4.3 shows two examples for particle shapes found after single-shot irradiation: Fig. 4.3a refers to intermediate intensity (around I_2), Fig. 4.3b to very high intensity (significantly above I_2). In the first case, a fairly spherical particle with a limited halo region is seen.

In contrast, at very high intensity there is, on one hand, a non-spherical central Ag particle but on the other hand a much larger region of small silver fragments. Considering that this image was taken after a single laser pulse only, it is quite plausible that after several pulses of sufficiently high intensity the particles are destroyed completely and the pertinent SPR band vanishes. In our experiments, total bleaching of the samples has been observed at intensities higher than 1.2 TW/cm^2 applying at least 5,000 pulses per spot. We interpret this finding as complete destruction of the Ag nanoparticles into small fragments without distinct SPR.

On the low-intensity side, in particular if one irradiates the sample with many pulses only slightly above the modification threshold (0.2–0.3 TW/cm^2), the maximum achievable spectral shift (and thus the maximum particle aspect ratio) is limited [6] because, due to successive particle deformation, the SP band polarized along the laser polarization moves out of resonance decreasing the interaction with the laser pulses. In the next section we will discuss how this problem can be circumvented by proper selection of irradiation wavelengths.

Summarizing this chapter, we want to point out that laser-induced shape transformation of silver nanoparticles embedded in glass using fs pulses requires a minimum peak pulse intensity of 0.2 TW/cm^2. Above this first threshold, linearly polarized pulses are able to create uniformly oriented, prolate spheroids with different aspect ratios depending on the actual intensity and number of pulses per spot.

Fig. 4.3 TEM of Ag nanoparticles in soda-lime glass after irradiation **a** in the region around $I_2 \approx 2$ TW/cm^2, **b** at significantly higher intensity $I > I_2$

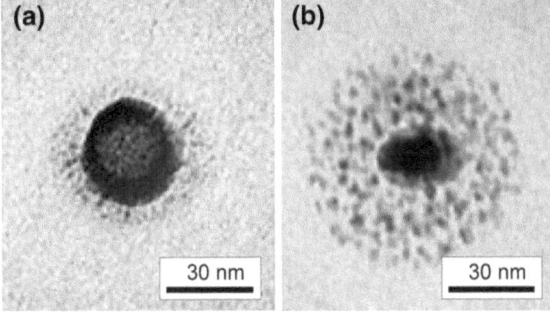

Exceeding a second threshold of ~ 2 TW/cm^2, one observes a reversal of the observed dichroism, which can be explained by oblate spheroids being produced by the laser pulses. In both cases the symmetry axes of the spheroids are oriented along the linear laser polarization. It should be also mentioned here that although oblate spheroids are also being created for circular polarization, their symmetry axes are given by the laser propagation direction [8] and such shape transformation is only possible in a low intensity, multi-shot mode. In all situations, too high an intensity or too many applied pulses cause destruction of the particles into very small fragments, macroscopically observable by fading of the SP absorption bands.

References

1. Podlipensky, A.V.: Laser assisted modification of optical and structural properties of composite glass with silver nanoparticles. Ph.D. Thesis, Martin-Luther-Universität Halle-Wittenberg. http://sundoc.bibliothek.uni-halle.de/dissonline/05/05H084/t1.pdf (2005)
2. Stalmashonak, A., Seifert, G., Graener, H.: Optical three-dimensional shape analysis of metallic nanoparticles after laser-induced deformation. Opt. Lett. **32**, 3215–3217 (2007)
3. Kaempfe, M., Rainer, T., Berg, K.-J., Seifert, G., Graener, H.: Ultrashort laser pulse induced deformation of silver nanoparticles in glass. Appl. Phys. Lett. **74**, 1200–1202 (1999)
4. Kaempfe, M., Seifert, G., Berg, K.-J., Hofmeister, H., Graener, H.: Polarization dependence of the permanent deformation of silver nanoparticles in glass by ultrashort laser pulses. Eur. Phys. J. D **16**, 237–240 (2001)
5. Seifert, G., Kaempfe, M., Berg, K.-J., Graener, H.: Production of "dichroitic" diffraction gratings in glasses containing silver nanoparticles via particle deformation with ultrashort laser pulses. Appl. Phys. B **73**, 355–359 (2001)
6. Podlipensky, A., Abdolvand, A., Seifert, G., Graener, H.: Femtosecond laser assisted production of dichroic 3D structures in composite glass containing Ag nanoparticles. Appl. Phys. A **80**, 1647–1652 (2005)
7. Podlipensky, A.V., Grebenev, V., Seifert, G., Graener, H.: Ionization and photomodification of Ag nanoparticles in soda-lime glass by 150 fs laser irradiation: a luminescence study. J. Lumin. **109**, 135–142 (2004)
8. Stalmashonak, A., Graener, H., Seifert, G.: Transformation of silver nanospheres embedded in glass to nanodisks using circularly polarized femtosecond pulses. Appl. Phys. Lett. **94**, 193111 (2009)

Chapter 5
"Off-Resonant" Excitation: Irradiation Wavelength Dependence

In this Chapter we will discuss the influence of various irradiation wavelengths on the induced dichroism in soda-lime glass with embedded silver nanoparticles. In the previous chapter we have seen that the position of the p-polarized SP band can only be red-shifted by a limited amount (to a peak position of 530 nm when using fs laser pulses at 400 nm). Any further increase in intensity or number of pulses applied to one sample position leads to bleaching of the SP bands, which can be explained by (partial) nanoparticle destruction. Many applications require polarization contrast at larger wavelengths in the visible and near IR spectral range; therefore, it is attractive to look for ways to meet these demands.

We will show that tuning of the irradiation wavelength is a very powerful parameter for reshaping Ag nanoparticles to large aspect ratios. Firstly, we have found that even rather strongly red-shifted excitation (with respect to the initial SP band) can, in spite of the low remaining SP absorption in this region, still very effectively induce a nanoparticle shape transformation to spheroids. In particular, such 'off-resonant' irradiation can create an even larger dichroism than resonant excitation. Secondly, we will demonstrate that subsequent irradiation by increasing laser wavelengths increases the particles' aspect ratio and thus the induced dichroism, allowing shifting of the p-polarized SP band further down to the near infrared region. Finally, we will show the results of simultaneous irradiation of the nanocomposites by pulses with different wavelengths, which result in a similar elongation of the nanoparticles, comparable to the ones obtained by subsequent irradiation.

5.1 Long Wavelength Irradiation

All of the experimental results reported below have been conducted in the multi-shot regime. Figure 5.1a gives the first example for "off-resonant" excitation. We will use this notion in the following to characterize a situation where the laser wavelength is considerably longer than the maximum SP resonance wavelength.

A. Stalmashonak et al., *Ultra-Short Pulsed Laser Engineered Metal–Glass Nanocomposites*, SpringerBriefs in Physics, DOI: 10.1007/978-3-319-00437-2_5, © The Author(s) 2013

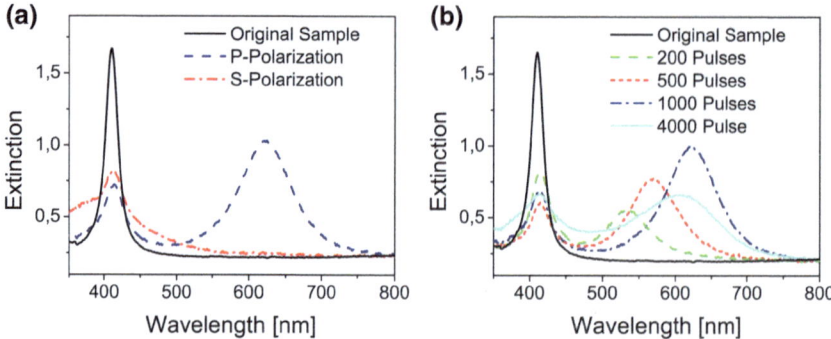

Fig. 5.1 Polarized extinction spectra of samples with Ag nanoparticles irradiated at 550 nm in the multi-shot regime. **a** 1,000 pulses in a single spot, peak pulse intensity 1.2 TW/cm^2. **b** P-Polarization, different numbers of pulses applied, peak pulse intensity 1.3 TW/cm^2

The polarized extinction spectra in Fig. 5.1a were measured on a sample containing Ag nanoparticles which was irradiated by 1,000 pulses per spot at $\lambda = 550$ nm with a peak pulse intensity of 1.2 TW/cm^2. The effect of this irradiation is similar to the one obtained by resonance excitation, namely, the original SP band of the spherical Ag nanoparticles peaked at $\lambda = 413$ nm splits into two polarization dependent bands. However, in this case, the p-polarized SP band (seen with light polarized parallel to the laser polarization) is peaked at 620 nm, while the s-polarization (perpendicular to laser) is shifted to shorter wavelengths, overlapped by a small residual absorption at 413 nm (due to the small, untransformed, particles [1]). This large spectral gap of the p- and s-polarized bands leads to a good polarization contrast at 620 nm, i.e., low and high transmission for p- and s-polarization, respectively. By analogy with the previous experiments, we can easily conclude that in the case of "off-resonant excitation" the nanoparticles are also transformed into prolate spheroids. From the spectral gap between the maxima of polarized extinction bands being significantly larger than in the case of resonant excitation, it is obvious that the aspect ratio of the reshaped nanoparticles is also larger than that of the resonant excitation.

Furthermore, similar to the case of resonant irradiation, the particle elongation and the corresponding magnitude of the induced dichroism can be tuned by variation of the peak pulse intensity and/or by the number of pulses per spot. Figure 5.1b illustrates the effect of different writing densities for samples irradiated at 550 nm with peak pulse intensity of 1.3 TW/cm^2. Only the p-polarized extinction spectra are shown. It is clearly seen that by increasing the number of applied pulses from 200 to 1,000, both peak wavelength and integrated extinction of the p-polarized band are increasing, while the absorption peak in s-polarization moves slightly towards shorter wavelengths (not shown in Fig. 5.1b). For a further increase of the writing density, however, the p-polarized SP band starts to shift back towards shorter wavelengths, accompanied by a decrease of amplitude and increase of bandwidth. This bleaching, which is due to the partial destruction of the silver nanoparticles, is clearly seen in the spectrum obtained after firing 4,000 pulses per spot into the sample. Once again, the off-resonant irradiation behaves very similarly to the resonant case: up to a certain number of pulses per spot, the dichroism can be increased by applying a larger number of pulses to the sample, but beyond this value (in the range of 1,000 pulses/spot) the maximal spectral shift is limited by the start of the nanoparticles' destruction process.

The results discussed so far show that irradiation of the samples at $\lambda = 550$ nm with optimum laser intensity and number of pulses leads to a larger spectral gap between the SP bands than the resonant excitation at $\lambda = 400$ nm. Therefore, one should expect a further increase of the induced dichroism when samples are being irradiated with even more off-resonant, longer wavelengths. We have also looked into this effect. Figure 5.2 shows three examples of extinction spectra for the samples irradiated at different wavelengths, here $\lambda = 490$ nm, 560 nm, and 610 nm. The parameters of the irradiation (laser intensity and number of pulses per spot) were chosen so that for each laser wavelength the maximum spectral shift was reached. It is clearly seen that in fact irradiation with the longer wavelengths leads to a larger spectral gap between polarized SP bands. However, as the irradiation wavelength moves further beyond 610 nm, the efficiency of the nanoparticles shape transformation decreases, and eventually the laser pulses no longer evoke any measurable extinction changes. For our samples this was the case for $\lambda \geq 670$ nm.

The obvious conclusion from the above results is that there is a threshold in absorption efficiency, which limits the long wavelength irradiation. This and all other findings are in good agreement with theory. As it was shown, the extinction efficiency decreases rapidly for wavelengths longer than the SP resonance and at 800 nm it becomes almost zero (see for example Fig. 3.6). At the same time, the E-field enhancement, which is present in the long wavelength region (Fig. 3.6), increases the probability of direct electron emission and makes the shape transformation of nanoparticles possible even for off-resonant excitation. As long as the

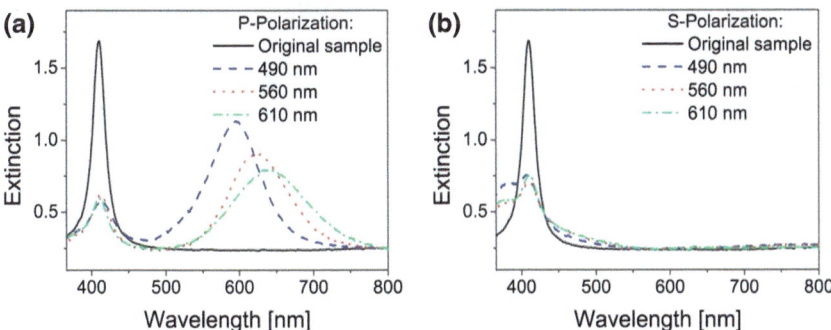

Fig. 5.2 Polarized extinction spectra of original and irradiated samples with Ag nanoparticles using different irradiation wavelengths. Intensity and number of pulses are optimized to achieve the best dichroism. **a** P-Polarization. **b** S-Polarization

absorption efficiency and laser intensity are high enough to emit ions from the particles and transform their shape, at least slightly after the first few pulses, the process is expected to work. This is due to the fact that even a minor change per pulse shifts the p-polarized SP band closer to resonance with the irradiation wavelength leading to a step by step increase in the efficiency of the shape transformation process. The process then will go on until the excitation wavelength is located considerably far into the blue wing of the p-polarized SP band. From then on the same mechanisms as in the case of resonant excitation leads to particle destruction and limits the spectral gap achievable by single-wavelength irradiation.

5.2 Subsequent Irradiation

As one could see in the previous section, long wavelength irradiation leads to a greater elongation of the nanoparticles. However, this type of irradiation has also its limits as in the case of resonance excitation. Nevertheless, we have found that this limitation can be overcome by multi-wavelength irradiation, i.e., subsequent irradiations of the same sample area by different laser wavelengths. If in particular after the first step the laser is tuned to another off-resonant position (on the longer-wavelength side of the already modified SP resonance) much larger dichroism as compared to the single-wavelength irradiation is observed. As an example, Fig. 5.3a shows the extinction spectra of a sample that was first irradiated at 535 nm, then at 670 nm with polarization parallel to the long axis of the already modified particles. It is clearly seen that the p-polarized band shifts in the second step further from the peak position of 560–760 nm. At the same time, the absorption peak in s-polarization shifts to shorter wavelengths. This large spectral gap between the s- and p-polarized bands corresponds to an aspect ratio of

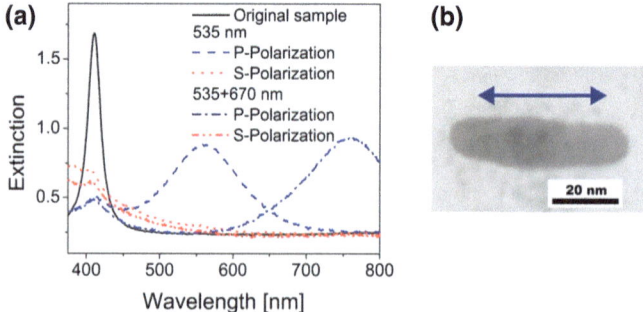

Fig. 5.3 **a** Polarized extinction spectra of samples with Ag nanoparticles irradiated firstly with 535 nm and subsequently with 670 nm laser pulses, 1,000 pulses per spot, peak pulse intensity was 1.5 TW/cm². **b** TEM image of transformed nanoparticles. Laser polarization is indicated by the *arrow*

(a/c) > 3 of the nanoparticles, which is proven by the TEM image shown in Fig. 5.3b. Clearly, subsequent irradiation with increasing laser wavelength leads to a very high degree of dichroism, and because of the minimal losses for the s-polarized light at 760 nm, this leads to a high degree of polarization contrast. It should be mentioned here that further red-shifting of the p-polarized SP band can be achieved by further irradiations with successively longer laser wavelengths at each step. We have performed preliminary experiments with a third irradiation at $\lambda = 800$ nm, which proved this idea. It is obvious that, by proper choice of the irradiation sequence and optimization of the pertinent laser parameters, the range of dichroism induced by this technique can be extended into the IR region.

5.3 Two-Wavelength Irradiation

If subsequent irradiation by different wavelengths can create larger aspect ratio nanoparticles embedded in the glass, it is an obvious idea to try simultaneous irradiation. We have done this and found that, other than in the previous case where the pulse intensities of subsequent irradiation were roughly the same (around 0.5–1.5 TW/cm^2), it is sufficient to include a small portion of the larger wavelength. In Fig. 5.4 we show the results of irradiating a sample simultaneously at 532 and 800 nm. The intensity of the green pulses was around 1.4 TW/cm^2, while the 800 nm pulses had an intensity that was lower by several orders of magnitude. Applying 1,000 pulses (for every wavelength) leads to similar results as presented above for the case of long wavelength irradiation (Fig. 5.4, dotted curve). The p-polarization band is peaked at ~ 570 nm, while the s-polarized SP band is shifted to the UV region (not shown). Increasing the number of pulses to 2,000 results in a bleaching (amplitude decrease) of the p-polarized SP band at ~ 570 nm and appearance of an additional band in the region of 750 nm (Fig. 5.4, blue, short-dashed curve). Further increase of the writing density strengthens this trend: the amplitude of the band located in the yellow region is decreasing with a

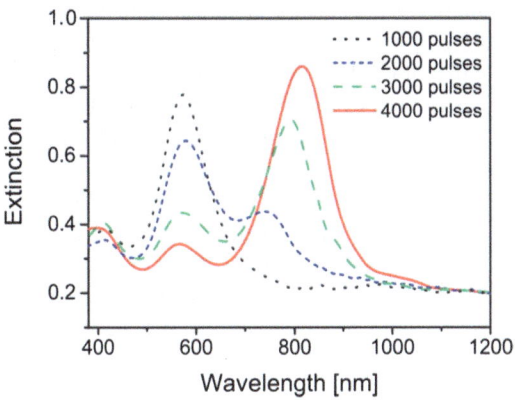

Fig. 5.4 P-polarized extinction spectra of a glass sample with Ag nanoparticles irradiated simultaneously by laser pulses at 532 and 800 nm

simultaneous shift of the SP resonance to shorter wavelengths, while the band located in the red region is increasing and shifts towards the IR region.

The obtained results can be explained in the following way: The first 1,000 pulses lead to the transformation of the initially spherical particles to prolate spheroids with an aspect ratio varying in the region of 2–2.5. Increasing the number of pulses, some of the nanoparticles with the highest aspect ratio start to elongate further. As a result, the sample contains a lower (with respect to the case of 1,000 irradiation pulses) number of particles with the aspect ratio of 2.5 and additionally some number of nanoparticles that have a greater elongation. From the spectra one observes that an amplitude decrease of the band corresponded to the NPs with a/c ~ 2.5 and an additional band for the elongated particles. Further increase of the writing density leads to a decrease in the number of nanoparticles with aspect ratio of 2.5 (and therefore, a decrease in amplitude) and further rise in the amount of longer NPs with a simultaneous increase of elongation (increase of amplitude and red-shift). The residual absorption at 550 nm for the case of irradiation by 4,000 pulses is due to the NPs with the lowest aspect ratio, which can not be transformed further [similar to the residual absorption at 413 nm in the case of off-resonant irradiation (e.g., Figs. 5.1 and 5.2).

However, the question of "how is it possible?" is still open. One of the possible and most probable effects, which can explain the obtained results, is the electric field enhancement and its dependence on wavelength. During irradiation by the first 1,000 pulses, the electric field enhancement created by the green pulses enhances the directed photoionization of the NPs and the existing processes are similar to the ones occurring for 'usual' irradiation, while the enhancement for the 800 nm pulses remains very weak. However, as the SP resonance band and the electric field enhancement factor are both shifted successively towards longer wavelengths, at some time after 1,000 (or more) pulses we reach a situation where the electric field enhancement factor is weak for the green pulses, while its value is now quite high for the pulses at 800 nm. Thus, the pulses at 532 nm are not very efficient for directed photoionization, but they are strong enough to excite the electrons. In turn, the weak pulses at 800 nm could not excite the nanoparticle, but the high EFE factor now enables directional ionization of the still very hot (since being excited by the green pulses) nanoparticle. As a result, the NPs are elongated further. It is also obvious that increasing the intensity of the IR pulses to the modification threshold can lead to the over-excitation of the NP, which will result in destruction of the latter. At the same time, if the intensities of both pulses will be lower that the modification threshold, then shape transformation will not be achieved.

In summary, we have to state that the laser-induced shape transformation of Ag nanoparticles is strongly dependent on the wavelength of the laser. The proposed technique of subsequent (or simultaneous) multi-wavelength, off-resonant irradiation of metal-glass nanocomposites has a huge potential for preparing polarizing elements with a high polarization contrast at any desired spectral position in the visible and near IR spectral ranges.

Reference

1. Stalmashonak, A., Seifert, G., Graener, H.: Spectral range extension of laser-induced dichroism in composite glass with silver nanoparticles. J. Opt. A: Pure Appl. Opt. **11**, 065001 (2009)

Chapter 6
The Effect of Temperature on the Laser-Induced Modifications of Ag Nanoparticles

In the previous section we have investigated the influence of laser irradiation parameters on the nanoparticles' shape transformation; now we focus on the effects of temperature during the reshaping processes. As discussed in Chap. 3, upon absorbing the energy of the laser pulse, the NP and its surrounding matrix experience a strong temperature increase of several hundred degrees for a time window of some ten picoseconds to a few nanoseconds. The pertinent softening of the adjacent glass matrix is crucial for the NP to have some degrees of freedom for the expansion and shape changes. On the other hand, it is known that heating the samples to temperatures above the glass transition temperature of soda-lime glass (~ 600 °C) is in any case restoring the spherical shape of the modified Ag nanoparticles [1]. This allows us to assume that the fs laser-assisted modification has to occur in the innermost shells where large transient, localized heating is present, while the outer shells of the surrounding glass should be cold enough to keep the anisotropic shape of the nanoparticle.

In addition to these effects, it has also been shown [2], that annealing at moderate temperatures after fs irradiation may cause modification of the SPR bands (and thus also of NP shape and the surrounding matrix). So, quite obviously the continuous global sample temperature, as well as the transient local heating and cooling, has considerable influence on the laser-induced shape modification of metallic NPs embedded in glass. Therefore, we will in the following discuss the influence of global heating of the samples as well as the local effects arising from heat accumulation in the focal volume of the fs laser irradiation as a function of laser repetition rate.

We start our discussion with the effect of irradiating a sample in a multi-shot regime with typical laser parameters, but keeping the sample temperature at a considerably elevated level during irradiation. Figure 6.1a shows polarized extinction spectra of a sample irradiated by 300 pulses per spot with a peak pulse intensity $I_p = 0.8$ TW/cm^2 at the laser wavelength $\lambda = 400$ nm at room temperature. In this case, the p-polarized SP band is centred at 507 nm while the s-polarization band is peaked in the UV region at 383 nm. However, irradiation of a preheated sample using the same laser parameters changes the extinction spectra (Fig. 6.1b). At a temperature of 125 °C, both SP bands are slightly shifted towards

A. Stalmashonak et al., *Ultra-Short Pulsed Laser Engineered Metal–Glass Nanocomposites*, SpringerBriefs in Physics, DOI: 10.1007/978-3-319-00437-2_6, © The Author(s) 2013

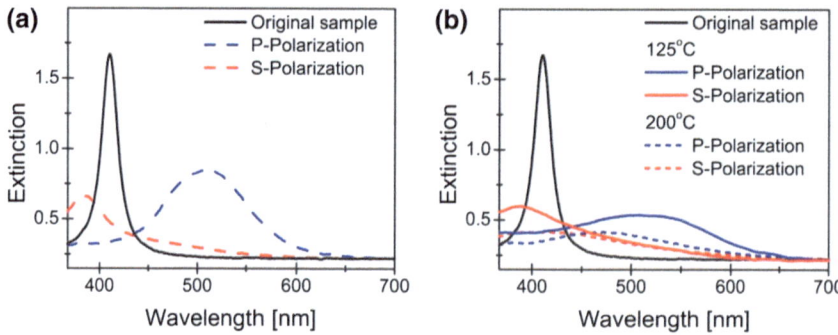

Fig. 6.1 Polarized extinction spectra of original and irradiated samples ($\lambda = 400$ nm, 300 pulses per spot, repetition rate 1 kHz, peak pulse intensity $I_p = 0.8$ TW/cm^2) **a** at room temperature, **b** at 125 and 200 °C

longer wavelengths and, in particular, are broadened and show considerably decreased amplitudes. A further increase of the global sample temperature to 200 °C results in almost complete bleaching of the plasmon bands (short-dashed lines in Fig. 6.1b). Concomitantly, the residual SP bands are further broadened, nearly making the induced dichroism disappear. So the first important experimental finding is that the relatively low temperature increase (with respect to the glass transition temperature) dramatically changes the results of the laser-induced NP shape modification. In order to gain a better understanding of the influence of temperature on the laser-induced modification of Ag NPs embedded in glass, we have measured the polarized spectra of samples irradiated at various temperatures from -100 to 170 °C in steps of 5–10 °C. The results in parameterized form are presented in Fig. 6.2. The spectral gap between the maxima of the polarized SP bands can be used as an approximate measure of the NPs' aspect ratio. The changes of the band integrals, which include the amplitude and bandwidth changes, are directly proportional to the absorption changes of the system.

Looking at Fig. 6.2a, first one can recognize three temperature intervals with different behaviors of the SP band centre positions. In the first interval from -100 to 80 °C the positions of the bands are nearly constant at ≈ 508 and ≈ 384 nm for the p- and s-polarization bands, respectively. Then, towards higher temperature (here up to ≈ 130 °C), the p-polarized SP band occurs at a red-shifted position ($\Delta\lambda_{max} \approx 20$ nm), while almost no shift of the s-polarized band is seen. Further increase of the sample temperature leads to a blue-shift of the p-polarized band with respect to the low-temperature limit. The behavior of the band integrals is given in Fig. 6.2b. Here again almost no temperature dependence is observed in the above-defined first interval (-100 to 80 °C), similar to the behavior of the band centres. At higher temperatures >80 °C, however, the band integrals of the p-polarized SP bands start to decrease very rapidly in amplitude. The amplitudes of the s-polarization bands decrease in a similar fashion, but at temperatures above ≈ 120 °C this effect is partially compensated by spectral broadening (compare

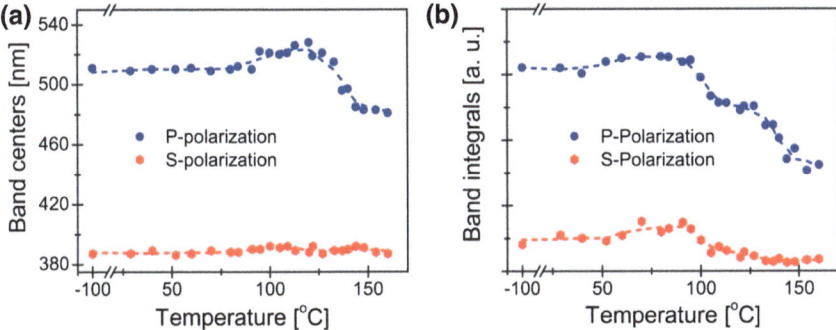

Fig. 6.2 Temperature dependence of polarized extinction spectra of **a** band centres, **b** band integrals

Fig. 6.1), which results in no further decrease of the band integrals above this temperature.

It should be mentioned here that we have also done similar series of irradiations with modified laser parameters (intensity and number of pulses applied); the results were comparable to those given above, i.e., the temperature dependence does not depend on the actual irradiation conditions. Interestingly, quite a similar behavior of the spectral changes discussed in Fig. 6.2 has been observed in a series of totally different experiments: the sample was irradiated at room temperature by fs pulses of different temporal separation, which was achieved by varying the laser repetition rate. Figure 6.3 shows selected results in parameterized form, i.e., the band centres (Fig. 6.3a) and the band integrals (Fig. 6.3b) derived from the polarized extinction spectra band as a function of the laser repetition rate. In Fig. 6.3a it is clearly seen that increasing the laser repetition rate from 1 to 10 kHz first leads to a small red-shift of the p-polarized band, but for rates ≥ 20 kHz both polarized SP bands are shifting back towards the original band of spherical nanoparticles. The corresponding band integrals are decreasing monotonously with

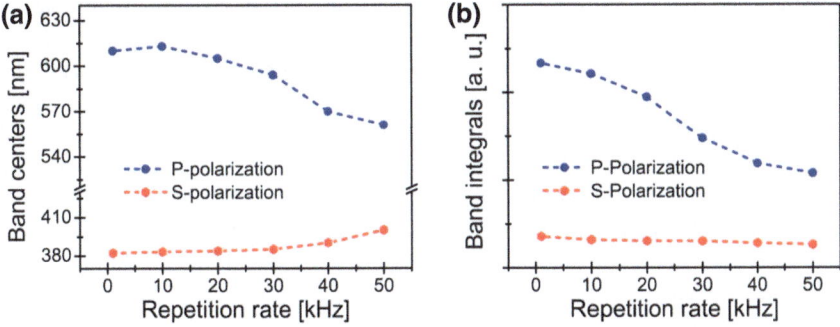

Fig. 6.3 **a** Dependence of polarized extinction spectra for band centres on repetition rate, **b** the corresponding dependence of band integrals

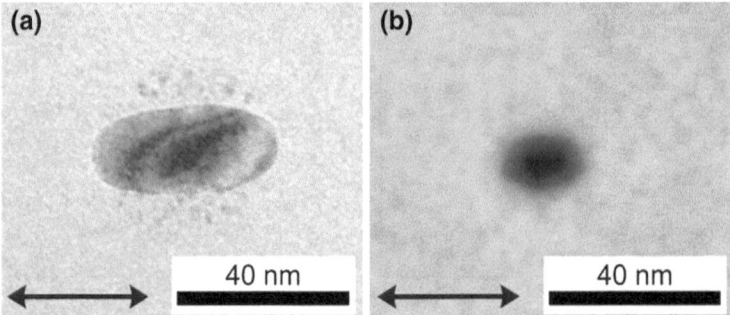

Fig. 6.4 TEM images of nanoparticles irradiated **a** at room temperature, **b** at 150 °C. Polarization of laser light is shown by an *arrow*

increasing repetition rates. At 100 kHz we have observed complete bleaching of the bands.

Regarding the physical mechanism of the NP shape transformation introduced above, we can conclude some novel details from the experimental findings shown in this chapter: first, for a globally increased matrix temperature the diffusion mobility of silver ions increases, leading to an enlarged radius of the cationic shell arising from Ag^+ emission after absorbing the fs laser pulse. Accordingly, the concentration of the silver cations in the immediate surroundings of the NP will decrease resulting in a lower precipitation rate. These processes are well suited to understanding the spectral changes observed when irradiating preheated samples.

Apparently, in the temperature range from -100 to 80 °C the ionic mobility is too low to cause significant diffusion of the emitted ions. Therefore, both the produced shapes of the nanoparticles (as an example see TEM image in Fig. 6.4a) and the pertinent SP bands are fairly constant in this regime. At slightly higher temperature, i.e., in the range of 100–120 °C, the optical extinction spectra start to show characteristic changes: the more prominent effect is the decrease of the band integrals. This can be understood by assuming that the mobility of Ag ions has now increased sufficiently such that the radius of the cationic shell grows; consequently the precipitation rate of silver atoms or clusters on to the main particle reduces, and the reduced volume of the resulting reshaped NP leads to a smaller optical absorption. The second effect, the moderate red-shift of both polarized SPR bands, can in this case be attributed to those Ag cations which are further away from the main NP, but still close enough to each other to precipitate as small clusters. These clusters will form an extended halo region that, via increasing the effective refractive index around the NP, can explain the observed small red-shift. Raising the temperature further leads to the increased mobility of the silver cations allowing them to diffuse so far away from the nanoparticle thereby diminishing their reduction and precipitation rates. Concomitantly, partial (and ultimately total) dissolution of the Ag NPs will occur instead of shape transformation to prolate spheroids. Figure 6.4b proves this assumption and shows an example of a partially dissolved NP. Additionally, this figure confirms that roughly 50 % of the NP volume is dissolved by ion emissions after applying 300 laser pulses. These experimental findings urge us to consider the effect of ion emissions on the NP-glass system as was introduced in Sect. 3.5.

Considering the above given calculations on the dynamics of heat flow, we can conclude that transient heating and cooling within a shell of 5–10 nm around the NP are crucial to the understanding of the shape change and dissolution processes. The results of the heat conduction calculations show that, starting from $T = 300$ K, the temperature at distances of ≥ 6–7 nm from the NP surface remains below 500 K at any time, while in the nearest shells of the matrix (distance from NP < 6–7 nm) high temperatures up to >1,000 K can be reached. Apparently this situation is required to promote the NP shape transformation on one hand, while protecting the nanoparticles from total dissolution on the other. Any change of parameters extending the spatial range around the NPs where temperatures clearly above 500 K occur, at least transiently, seems to enable particle dissolution. This holds for the temperature-dependent studies in this work as well as for experiments with considerably higher laser intensity, which also resulted in partial dissolution of the NPs.

If this picture is correct, it should also provide an explanation for the results observed by irradiation of NPs by laser pulses with different repetition rates, i.e., we have to look for a connection between the temporal separation of the laser pulses and the local sample temperature. The connection can be found by considering the heat flow from the focal volume (which in the beam direction is limited to a thickness of ≈ 2 μm by the layer containing the NPs) to the cold parts of the sample. The total thickness of the glass substrate is 1 mm. Using these parameters and Eq. 3.6 (which describes the heat transfer), we have calculated the temperature accumulation in the focal volume as a function of pulse repetition rate and number of pulses applied. The pertinent temperature rise ΔT in the focal volume as a function of number of pulses is shown in Fig. 6.5a. It is clearly seen that every pulse increases the temperature in the focal volume. Applying 300 pulses to one spot with temporal separation of 1 ms (1 kHz repetition rate) results

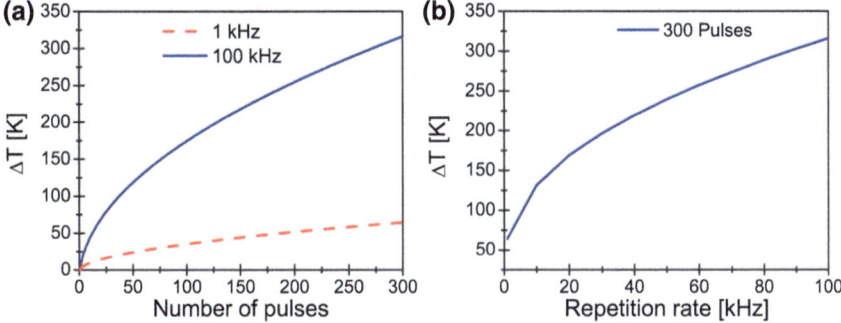

Fig. 6.5 Increase of the temperature in the focal volume as a function of **a** applied number of pulses for the cases of irradiation at 1 and 100 kHz, **b** repetition rate when applying 300 pulses

in a temperature rise of about 50–60 K, while the same number of pulses at 100 kHz repetition rate increases the temperature by more than 300 K. Figure 6.5b shows the dependence on the laser repetition rate of ΔT after 300 pulses.

These simulations are in very good agreement with our experimental results: Fig. 6.5b indicates that irradiation of nanoparticles by 300 pulses at repetition rates below 10 kHz leads to a relatively small increase of temperature in the focal volume (<130 K for 300th pulse). As shown in Fig. 6.2 this is typically the constant temperature where NP dissolution begins. For repetition rates above 10 kHz this region of $\Delta T > 100$ K is reached increasingly earlier, which should cause a reduced spectral gap and higher degree of NP dissolution. Just this behavior is seen in the experimentally obtained dependence on laser repetition rate shown in Fig. 6.3. Finally, the observation that a higher number of pulses allows the dissolution process to appear at lower repetition rates is nicely compatible with this calculation, since, e.g., in the case of 600 pulses the first 300 prepare a considerable temperature rise, ΔT, which enables dissolution by the 'second half' of the pulse train.

In conclusion, we have shown that the intended transformation of the initially spherical NPs to prolate shapes with high aspect ratio by irradiation with a few hundred laser pulses requires a special spatial-temporal evolution of the matrix temperature: the heat-affected zone reaching transiently temperatures above ≈ 500 K around a nanoparticle should apparently be limited to a few nanometers. Then the increased mobility of the emitted Ag ions enables the processes of shape transformation within that shell, while the cooler outer shells prevent the ions from drifting farther away from the NP. The latter obviously happens when the initial sample temperature lies above 100 °C, which can also be reached by accumulation of the absorbed energy in the focal volume when applying high laser repetition rates. In this case the emitted Ag cations seem to drift so far away from the NP that they cannot diffuse back to the NP and recombine with it, but rather precipitate where they are, forming an enlarged halo region. This process readily explains the observed dissolution of the NPs after irradiation with a few hundred fs laser pulses. It should be mentioned that for different irradiation parameters (intensity, focal size, number of pulses per spot) the temperature accumulation effects can be minimized, so that successful reshaping is possible up to repetition rates of >100 kHz.

References

1. Podlipensky, A., Abdolvand, A., Seifert, G., Graener. H.: Femtosecond laser assisted production of dichroic 3D structures in composite glass containing Ag nanoparticles. Appl. Phys. A **80**, 1647–1652 (2005)
2. Podlipensky, A.V., Grebenev, V., Seifert, G., Graener, H.: Ionization and photomodification of Ag nanoparticles in soda-lime glass by 150 fs laser irradiation: a luminescence study. J. Lumin. **109**, 135–142 (2004)

Chapter 7
Ultra-Short Pulsed Laser Engineering of Metal–Glass Nanocomposites

7.1 Fabrication of Sub-Micron Polarizing Structures

It was shown in the previous sections that the shapes of the initially spherical Ag NPs in glass can be permanently modified by femtosecond laser irradiation; that in turn leads to the optical dichroism induced in the composite glass. The induced degree of dichroism (spectral gap between the two polarized resonances), being correlated with the aspect ratio (ratio of long and short axis) of the NPs, depends on the irradiation parameters such as wavelength, peak intensity and number of laser pulses applied. We will now explore the fabrication of polarization and wavelength selective micro-devices based on metal-glass nanocomposites. To achieve a considerable degree of polarization contrast for such polarizing elements, the total extinction of the shape-transformed NPs must be correspondingly high. This obviously requires the initial nanocomposite materials to have a high concentration of NPs. Under such conditions, it may become necessary for several successive irradiations to be performed in order to achieve the ultimately desired spectral parameters. This should be followed by further annealing of the samples to remove colour centres and other defects in the glass matrix. We have recently studied a number of technical points to set the framework for further optimization [1]. As this requires usually a complex interplay of finding the best combination of sample (NP sizes, their spatial distribution etc.,) and irradiation parameters, we refrain from discussing here all the details of this optimization task. Instead, we will only give some instructive examples for polarizing optical micron structures that we have already demonstrated to exhibit the potential for a number of applications.

Micro-polarizing structures fabricated in composite glass containing silver nanoparticles are shown in Fig. 7.1. In order to achieve spots with sizes of a few micrometers, one requires focusing with a lens of sufficiently high numerical aperture. In this experiment the laser beam diameter was first enlarged to 15 mm and then focused with a lens of either 55 or 80 mm focal length.

We have irradiated the sample using pulses at $\lambda = 515$ nm; the laser parameters were chosen so that maximum dichroism was obtained. The matrix seen in

A. Stalmashonak et al., *Ultra-Short Pulsed Laser Engineered*
Metal–Glass Nanocomposites, SpringerBriefs in Physics,
DOI: 10.1007/978-3-319-00437-2_7, © The Author(s) 2013

Fig. 7.1 Polarized matrix made by laser irradiation of silver nanoparticles embedded in glass. Polarization of the light is parallel to the laser polarization. If the light is polarized perpendicularly to the laser polarization, the spots are disappearing. The size of one spot is around 3 μm

Fig. 7.1 consists of spots (modified area) with a size of around 3 μm and a distance of 15 μm between them. The spots are seen when polarization of the light (in a microscope) is parallel to the laser polarization. If the light is polarized perpendicularly to the laser polarization, the spots disappear. The irradiated areas have a green colour in p-polarization because of the extinction at around 600 nm.

The next example is shown in Fig. 7.2. Here, the irradiation parameters are the same for every spot. However, the pattern of the structure, in this case, consists of three irradiated and one non-irradiated spots and the laser polarization is rotated on 45° and 90° for every next irradiated spot in the pattern. The four photos shown in Fig. 7.2 were taken from the same area on the sample. The only difference is the polarization of light in the microscope, which for every case is shown as an arrow. The colour of the spots is darker than in the previous case, because the irradiated sample had a higher filling factor, which leads to the broader extinction.

The last example presented in Fig. 7.3 shows a similar structure to that shown in Fig. 7.2. Again, the pattern consists of three irradiated and one non-irradiated spots and the laser polarization is rotated on 45° and 90° for every next irradiated spot in the pattern. However, in this case the irradiation parameters for every spot were different. The increase in the number of pulses leads to the red-shift of the p-polarized SP band and as a result the colour is changed from brown to green. On the other hand, rotation of the probe light polarization leads to the changes of the spots' colours, so that the spot that has a red colour in one polarization becomes yellow–brown for the light polarized in the perpendicular direction.

In summary, we have managed to prepare several regular micro-patterns with tailored dichroism (with a spot size of about 3 μm) by laser irradiation of nano-composite glass with embedded silver nanoparticles. These and any other

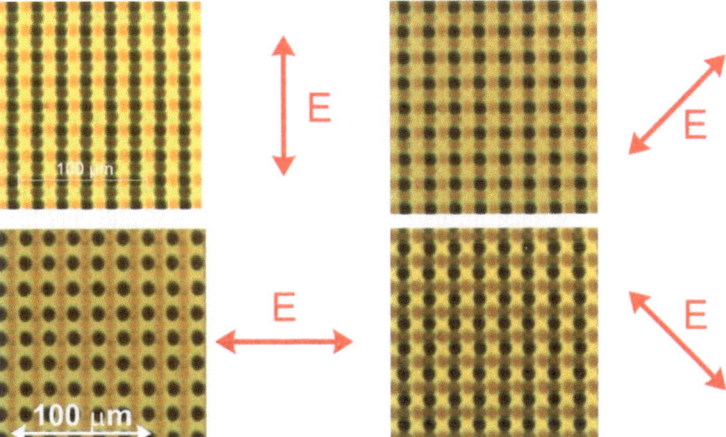

Fig. 7.2 Polarized microstructure made by laser irradiation of silver nanoparticles embedded in glass. Parameters of irradiation (except laser polarization) are the same for every spot. Polarization of the light forming the microscope images is shown by the *red arrows*

microstructures are not subject to any mechanical limitations, which would have been the case if similar polarization patterns had to be produce by standard technology (i.e., cutting and pasting of large polarizers); only the diffraction limit sets a lower size limit of a few hundred nanometers for the structures that can be created. Consequently the technology is well suited to produce polarization and wavelength selective micro-devices, allowing unprecedented feature sizes currently not achievable by any other method.

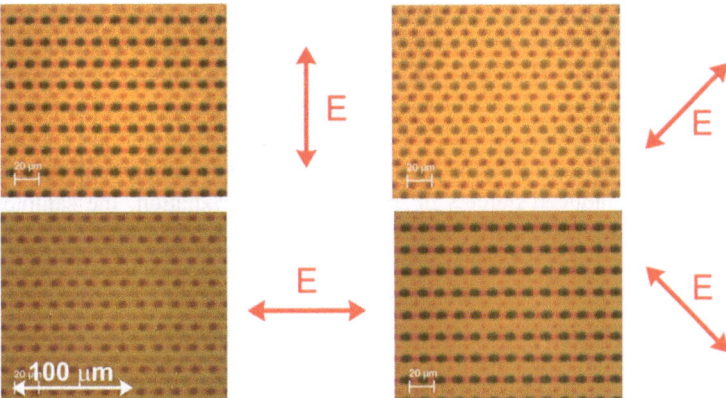

Fig. 7.3 Polarized microstructure made by laser irradiation of silver nanoparticles embedded in glass. Parameters of irradiation are different for every spot in the pattern. Polarization of the light forming the microscope images is shown by the *red arrows*

7.2 Metal–Glass Nanocomposites for Optical Storage of Information

Recently, besides continuous improvements to the established materials and technologies for data storage, several new concepts in this field have been reported. While most of these studies have been aiming at higher capacity or access speed of the storage media (e.g., [2, 3]), only a few attempts have been made towards materials promising durability for centuries [4, 5]. Among such approaches are ideas such as nano-objects being moved electromechanically within carbon nanotubes [6], which are tentatively still decades away from their practical realization, and an interesting all-optical data storage technique based on silver clusters (Ag_m^{x+}, m < 10) in glass [4]. In the latter case, the Ag clusters are formed within the focal volume of each femtosecond pulsed laser irradiated zone; photoluminescence from this volume is then used for data readout.

We present an attractive approach to optically store on, and read data from, glass with embedded silver nanoparticles of proven long-term stability. The recording technique is once again based on femtosecond pulsed laser-induced shape transformation of the initially spherical nanoparticles into prolate (cigar-like) spheroids. The rotational (longer) axes of these nano-objects are uniformly orientated along the polarization direction of the linearly polarized laser pulses. We demonstrate the capabilities of this effect for storing data in the absorption characteristics of each irradiated spot, which can be as small as the diffraction limit permits. Prospectively, and with the help of near-field enhancement techniques, several bits of information could be stored in the thermodynamically stable shape of even a single nanoparticle. The technology has the potential for data capacity and access speeds superior to the currently available optical recording media, combined with unprecedented durability.

The nanocomposites were irradiated using a second harmonic beam at 515 nm from a pulsed Yb:KGW laser, generating 300-fs pulses at 100 kHz pulse repetition rate. The beam was focused to a spot with near-Gaussian intensity profile and a full width half maximum (FWHM) of approximately 0.5 μm using a microscope objective with a numerical aperture (N.A.) of 0.9. The samples were mounted on a motorized X–Y microscope translation stage. The maximum laser peak pulse intensity used was ∼2 TW/cm^2. In order to remove colour centres and other metastable laser-induced defects in the glass after irradiation, the samples were annealed for 1 h at 200 °C.

Several hundred pulses in an energy density range of 20–50 mJ/cm^2 lead to well-defined dichroic changes in the local absorption spectrum of the sample. The initially isotropic surface plasmon absorption band of the nanocomposite, peaked typically at 410 nm, is split into one slightly blue-shifted band (around 390 nm, seen with polarization perpendicular to the laser), and a considerably broadened and red-shifted band (in the range of 500–650 nm, seen with polarization parallel to the laser). The position of the bands is controlled by the peak intensity and the number of pulses fired per spot. The red-shifted band is chosen for data storage applications, because it does not exhibit any spectral overlap with residual absorption from spherical nanoparticles, and thus extinction spectra with high polarization contrast can be produced at several positions within this spectral range (from 500 to 650 nm).

In order to demonstrate that our technique works down to the diffraction limit, we show dichroic spots, each ~ 500 nm in size, with the colour under polarized illumination being determined by the number of pulses fired onto each spot (Fig. 7.4a). This figure shows examples of the diffraction-limited dichroic spots written at a 1 µm pitch. Four columns have been written in the sample. For irradiation of each column the same linear polarization of the laser beam was used. However, looking at the four columns and from left to right, the polarization axis of the laser beam was rotated by 0, 45, 135 and 90° with respect to a horizontal axis. The individual five spots in every column were prepared by changing the number of pulses fired onto each spot, increasing from top (100 pulses) to bottom (500 pulses). The four images in Fig. 7.4a, with readout being performed with a different orientation of the polarizer in front of the camera ("Olympus" microscope with uEye CCD Camera), clearly show that dark spots of different colours in one column can be made almost invisible by simple rotation of the light polarization.

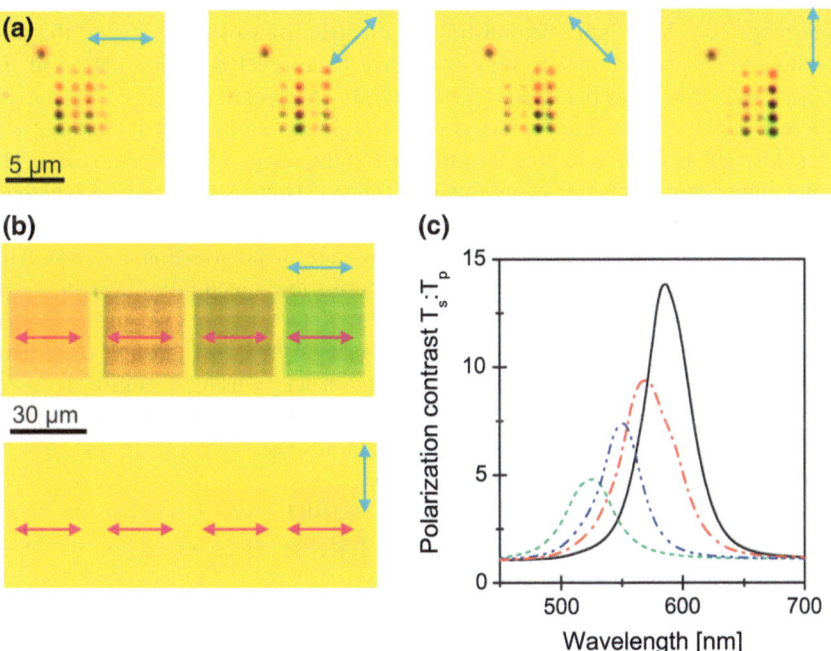

Fig. 7.4 **a** Examples of ~ 500 nm dichroic spots written by laser into the sample. *Blue arrows* show the orientations of the polarizer in front of the camera used for readout of the images. **b** Examples of large dichroic areas. From *left* to *right*, the number of pulses per spot was increased from 200 to 500. Here, the *red arrows* indicate the polarization of laser light during writing. **c** The polarization contrast as a function of wavelength. The figure is reprinted with permission from [7]. Copyright (2011), American Institute of Physics

To analyse the spectral features of the irradiated spots with satisfactory signal-to-noise ratio, we measured the absorption spectra (microscope spectrophotometer MPM 800 D/UV, Zeiss) on somewhat larger areas (30 × 30 μm^2) irradiated with the same parameters as the single, diffraction-limited spots (Fig. 7.4b). This figure gives four examples of such squares, where the red arrows indicate the polarization of laser light during writing, and the blue arrows the polarization of the light for recording the images. From left to right, the number of pulses per spot fired onto the sample was increased from 200 to 500. The polarization contrast—defined as the ratio of the transmitted intensity under polarized illumination (*Ts-wave/Tp-wave*)—as a function of wavelength is also presented (Fig. 7.4c). The spectra show that position of the absorption band can be controlled by the laser parameters. The polarization splitting of the surface plasmon absorption bands, measured with light polarized along the short and long axes of the spheroids, yields a polarization contrast of ∼ 10 for the individually tailored maximum wavelengths of 525, 550, 567 and 585 nm. Thus, up to four bits of information could be encoded via this wavelength variation technique.

In addition to this, the polarization axis of the laser light is another easily controllable parameter for encoding several bits of information; since the transmission of linearly polarized light depends monotonously on the angle between the dichroism inscribed in the nanocomposite and the polarization axis of the readout light (Fig. 7.5). This figure demonstrates the capabilities of this concept: the nine green–blue squares, shown in Fig. 7.5a, have been created by irradiation of the sample with the same parameters in terms of wavelength and intensity but for different angles of the laser polarization; the linear polarization of the laser was rotated by 10° from area to area (as indicated in the image by the red arrows). The visual impression makes it clear that the transmission of light at a wavelength within the surface plasmon absorption band of the long nanoparticle axis can be altered (gradually increased or decreased) by careful polarization rotation. This effect is demonstrated in detail in Fig. 7.6, where a few spectra for different values of the polarization angle are shown (Fig. 7.6a). In addition, we measured the transmission of a linearly polarized He–Ne laser (at 633 nm) through one of the irradiated areas as a function of the polarization angle using a confocal microscope. As can be seen from Fig. 7.6b, a simple transmission measurement using a monochromatic light source can easily encode at least eight values, i.e., a 3-bit word, for appropriately chosen relative angles.

Adding up wavelength and polarization variation, we suggest the following approach for a reliable and fast simultaneous readout of the stored data—as shown schematically in Fig. 7.7. Four beams of circularly polarized monochromatic light of different colours, which correspond to the maxima of the extinction bands (for example for Fig. 7.4c—525, 550, 567 and 585 nm), can be focused to the same spot on the storage medium. Passing through the spot, each beam is being partially (elliptically) polarized with respect to the extinction band of this spot. The largest proportion of the polarized light (highest polarization contrast) will be associated with the beam whose colour matches the maximum (peak) of the extinction band. As an example we can assume that the spot has the optical (extinction) properties corresponding to the red band in Fig. 7.4c.

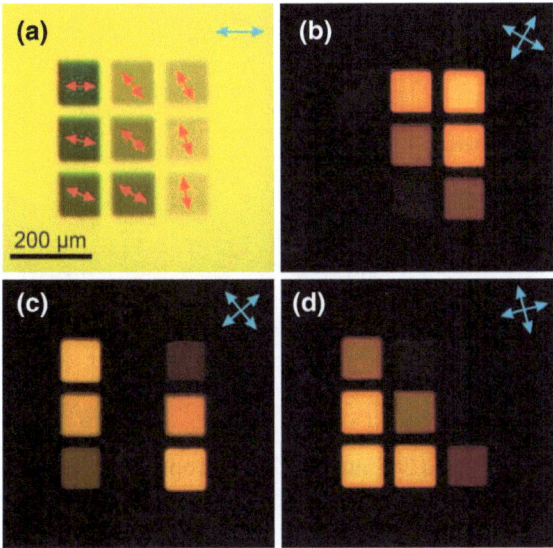

Fig. 7.5 *Nine squares* are written by irradiating the sample with the same laser parameters (intensity and number of pulses per spot fired) but for different linear laser polarizations. *Red arrows* show the direction of the laser polarization irradiating each area. **a** The image is taken with one polarizer in front of the camera. The *blue arrow* defines the direction of polarization for this polarizer. **b, d** Here, the images are taken when the sample is placed between two cross polarizers. *Blue arrows* define the polarization direction of the polarizers (**b**) polarizers are rotated by 20° with respect to the sample, (**c**) 45°, and in (**d**) 70°. The figure is reprinted with permission from [7]. Copyright (2011), American Institute of Physics

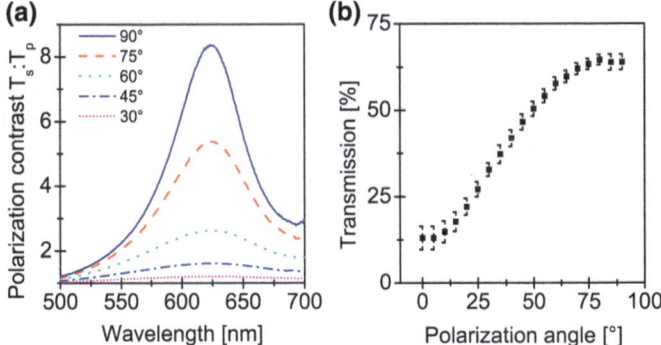

Fig. 7.6 **a** Spectra of polarization contrast for different values of angle between the dichroism inscribed in the nanocomposite and polarization axis of the light. **b** Transmission of a linearly polarized He–Ne laser through the irradiated area as a function of the polarization angle. The figure is reprinted with permission from [7]. Copyright (2011), American Institute of Physics

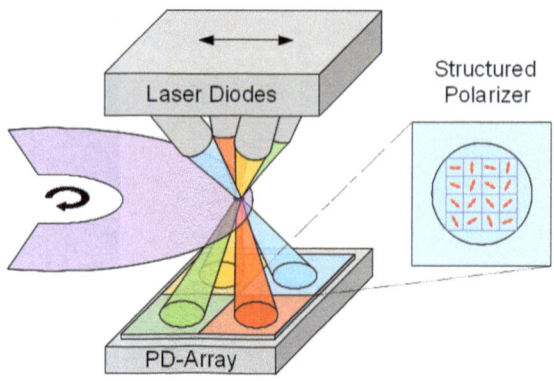

Fig. 7.7 Proposed readout approach. *Red arrows* show the polarization axis of the structured polarizer; PD-Array—an array of photo-diodes. The figure is reprinted with permission from [7]. Copyright (2011), American Institute of Physics

When passing through the spot, the 525 nm beam will have a polarization contrast of $\sim 2.5{:}1$, 550 and 585 nm—$\sim 7{:}1$, and only 567 nm will have the highest contrast of $\sim 9{:}1$. The next step would then be to determine the polarization direction or axis of the already elliptically polarized light. This can be done with the help of tailored structured polarizers (made using a similar technology to the data storage), which consist of eight pairs of "90°-polarizers" rotated at 10°–11° relative to each other (Fig. 7.7). The transmitted light will then fall onto the PD-Array—an array of photo-diodes—for registration. The highest "polarization contrast" (the ratio of transmitted intensities for each pair of "90°-polarizers") (9:1 for our example) will be achieved only for that pair of "90°-polarizers", whose polarization axis matches the polarization axis of the transmitted light through the storage medium. Having such structured polarizers for each beam will allow us to unambiguously determine the wavelength and relative orientation of the highest polarization contrast simultaneously (i.e., in one-step).

Storage capacity, writing and readout speeds achievable with the proposed long-term storage medium are then estimated as follows: as supposed above, assuming four different wavelengths and eight polarization states, we can encode five bits (32 values) on to one spot. Together with the diffraction-limited spot diameter of 600 nm (the longest wavelength used for readout) this provides a storage capacity of ~ 30 GB for a DVD-like one-layer disc, i.e., the same range as the state-of-the-art Blu-ray technology offers. Typical writing of information requires sub-picosecond laser pulses with an energy of ≤ 1.5 nJ per spot; nowadays fibre lasers with 1 W of average power at 515 nm are commercially available, apparently allowing for a writing speeds of 5 Mbit/s, which is comparable to the DVD technology standard. Readout speed is only limited by the response time of the photodetectors used; thus, regarding the 5 bits per spot, readout rates of 100 Mbit/s and more may realistically be expected, which are superior to the current DVD and Blu-ray technologies.

In summary, we have presented an approach to optically store on, and read data from, glass with embedded silver nanoparticles. Prospectively, and with the help of near-field enhancement techniques, several bits of information could be stored in the thermodynamically stable shape of even a single nanoparticle. The technology has the potential for data capacity and access speeds superior to the currently available optical recording media, combined with unprecedented durability. The durability of the storage medium is basically that of the glass itself, i.e., it is not changed by temperatures up to 600 °C, and is insensitive to the most chemical agents and biological ageing.

References

1. Stalmashonak, A., Seifert, G., Unal, A.A., Skrzypczak, U., Podlipensky, A., Abdolvand, A., Graener, H.: Toward the production of micropolarizers by irradiation of composite glasses with silver nanoparticles. Appl. Opt. **48**, F38–F44 (2009)
2. Oslon, C.E., Prvite, J.R., Fourkas, J.T.: Efficient and robust multiphoton data storage in molecular glasses and highly crosslinked polymers. Nature Mater. **1**, 225–228 (2002)
3. Zijlstra, P., Chon, J.W.M., Gu, M.: Five-dimensional optical recording mediated by surface plasmons in gold nanorods. Nature **459**, 410–413 (2009)
4. Begtrup, G.E., Gannett, W., Yuzvinsky, T.D., Crespi, Y.H., Zettl, A.: Nanoscale reversible mass transport for archival memory. Nano. Lett. **9**, 1835–1838 (2009)
5. Royon, A., Bourhis, K., Bellec, M., Papon, G., Bousquet, B., Deshayes, Y., Cardinal, T., Canioni, L.: Silver clusters embedded in glass as a perennial high capacity optical recording medium. Adv. Mater. **22**, 5282–5286 (2010)
6. Shimotsuma, Y., Sakakura, M., Kazansky, P.G., Beresna, M., Qiu, J., Miura, K., Hirao, K.: Ultrafast manipulation of Self-Assembled form birefringence in glass. Adv. Mater. **22**, 4039–4043 (2010)
7. Stalmashonak, A., Abdolvand, A., Seifert, G.: Metal-glass nanocomposite for optical storage of information. Appl. Phys. Lett. **99**, 201904 (2011)

Chapter 8
Conclusion

The experimental results presented in this brief are mostly concentrated on the investigation of the laser-induced shape modification of initially spherical silver nanoparticles incorporated in glass and the processes leading to the different NP shapes. A detailed understanding of all these processes helps to optimize and establish this technique and allow us to modify the optical properties of composite glasses containing metal nanoparticles for applications in photonics.

The dependence of the nanoparticle shapes on the laser polarization was discussed. It was found that irradiation of the NPs by laser pulses with circular polarization leads to the transformation of initially spherical particles to oblate spheroids with their symmetry axis parallel to the propagation direction. In the case of linearly polarized light, however, the NP shape can be modified to either prolate or oblate spheroids with symmetry axes parallel to the laser polarization. The shape in this case is defined by the laser peak pulse intensity used for irradiation. Pulse intensities slightly above a first modification threshold lead to the elongation of the silver nanoparticle parallel to the laser polarization (prolate spheroid). On the other hand, increasing the peak pulse intensity by one order of magnitude above a second threshold results in oblate spheroids but, in this case, with the short axis parallel to the laser polarization. These results allowed us to conclude that the main process responsible for the different NPs' shape transformation is the directional photoionization.

We have also found that laser-induced shape transformation of Ag nanoparticles is strongly dependent on the laser wavelength. The first striking observation is that considerably off-resonant excitation [i.e., irradiation with a laser wavelength shifted more than 100 nm to the longer wavelength side of the SP resonance absorption of spherical nanoparticles (at 413 nm)] can even more effectively transform the shapes of the nanoparticles to spheroids with large aspect ratios than near-resonant interaction, despite the very weak coupling to the SPR in this region.

The fact that the laser-assisted elongation of nanoparticles stops when the excitation wavelength is located considerably far into the blue wing of the p-polarized SP band, together with the results obtained in experiments with simultaneous irradiation of the sample by two wavelengths, allowed us to conclude

A. Stalmashonak et al., *Ultra-Short Pulsed Laser Engineered Metal–Glass Nanocomposites*, SpringerBriefs in Physics, DOI: 10.1007/978-3-319-00437-2_8, © The Author(s) 2013

that a very weak electric field enhancement at the wavelength of irradiation can be the main factor limiting the laser-induced dichroism. This limit can be overcome by subsequent irradiation, tuning the irradiation pulses to the longer wavelengths.

Taking into account the theoretical estimations and experimental results observed by the study of temperature dependence we can conclude that the intended transformation of initially spherical NPs to prolate shapes with high aspect ratio by irradiation with a few hundred laser pulses requires a special spatial-temporal evolution of the matrix temperature: the extent of the heat-affected zone reaching transiently temperatures above ≈ 500 K around a nano-particle should apparently be limited to a few nanometers.

According to all the acquired experimental results and calculations we also proposed the possible deformation mechanisms based on the transient phenomena that are controlled either directly by the electric field of the laser pulse or indirectly by the temperature rise induced by it. Formation of the prolate spheroids with their long axis parallel to the laser polarization in the low intensity range for multi-shot irradiation could be explained by a combination of the photoionization and metal particle precipitation on the poles of the nanosphere. The intensity-dependent extension of the cationic shell around the nanoparticle and the photoelectron emission in the direction of the laser polarization play a key role here. In the case of high intensities (above 2 TW/cm^2) and low number of pulses (less than 40), dense electron plasma formation at the poles of the sphere and subsequent thermal expansion or even ablation of the glass matrix dominate, leading to the transformation of nano-spheres to oblate spheroids.

All these findings allowed us to optimize the laser-assisted modification technique, manipulating and engineering the optical properties of metal-glass nanocomposites. This led to the fabrication of (sub-) micro-polarizing structures (polarization and wavelength selective devices) with high polarization contrast and a broad tunable range of dichroism, and also to optical storage of information in the nanocomposites.